RETROVIRAL PR

Control of maturation an

RETROVIRAL PROTEASES

Control of maturation and morphogenesis

Edited by

Laurence H. Pearl

Department of Biochemistry
University College, London

First published 1990

Published by
THE MACMILLAN PRESS LTD
Houndmills, Basingstoke, Hampshire RG21 2XS
and London
Companies and representatives
throughout the world

Published in the United States and Canada by
STOCKTON PRESS
15 East 26th Street, New York, NY 10010

Printed in Great Britain by Billings, Worcester

British Library Cataloguing in Publication Data
Retroviral proteases: control of maturation and morphogenesis.
1. Man. Proteinases. Biochemistry
I. Pearl, Laurence
612.01516
ISBN 0–333–53612–6

Library of Congress Cataloging-in-Publication Data
Retroviral proteases: control of maturation and morphogenesis/
edited by Laurence H. Pearl
p. cm.
Proceedings of a meeting held in York, England, Dec. 4–5, 1989 and
sponsored by the Medical Research Council AIDS Directed Programme.
ISBN 0–935859–95–0: $45.00
1. Proteolytic enzymes—Congresses. 2. Retroviruses—Enzymes—
Congresses. 3. Proteolytic enzyme genes—Congresses. 4. HIV
(Viruses)—Enzymes—Congresses. 5. Viral genetics—Congresses.
I. Pearl, Laurence H., 1956– . II. Medical Research Council
(Great Britain). AIDS Directed Programme.
[DNLM: 1. HIV—Enzymology—Congresses. 2. Peptide Hydrolases—
Metabolism—Congresses. 3. Retroviridae—Enzymology—Congresses.
QW 166 R4355 1989]
QP609.P78R47 1990
616.0194—dc20
DNLM/DCL
for Library of Congress

90–9413
CIP

Contents

The Contributors

Tom L. Blundell
Laboratory of Molecular Biology, Department of Crystallography, Birkbeck College, Malet Street, London WC1E 7HX, London, U.K.

Ming Bu
Department of Microbiology, Immunology and Parasitology, Louisiana State University Medical Center, New Orleans, LA, U.S.A.

K.-H. Budt
Hoechst AG, Pharma Forschung Allgemein, D-6230 Frankfurt 80, F.R.G.

Peter Calkins
Department of Microbiology, Immunology and Parasitology, Louisiana State University Medical Center, New Orleans, LA, U.S.A.

Nancy T. Chang
Tanox Biosystems Inc., 10301 Stalla Link, Houston, Texas 77025, U.S.A.

Tse Wen Chang
Tanox Biosystems Inc., 10301 Stalla Link, Houston, Texas 77025, U.S.A.

Robert J. Craig
Smith Kline Beecham Pharamaceuticals, Research Laboratories, King of Prussia, Pennsylvania 19406, U.S.A.

Dennis E. Danley
Pfizer Central Research, Eastern Point Road, Groton, Connecticut 06340, U.S.A.

Christine Debouck
Smith Kline Beecham Pharamaceuticals, Research Laboratories, King of Prussia, Pennsylvania 19406, U.S.A.

Ingrid C. Deckman
Smith Kline Beecham Pharamaceuticals, Research Laboratories, King of Prussia, Pennsylvania 19406, U.S.A.

Albert G. Dee
Division of Molecular Biology, Lilly Research Laboratories, Indianapolis, Indiana 46285, U.S.A.

Duncan Gaskin
Laboratory of Molecular Biology, Department of Crystallography, Birkbeck College, Malet Street, London WC1E 7HX, London, U.K.

Hans Gelderblom
Institute of Virology, Robert Koch-Institute of the Federal Health Administration, D-1000 Berlin 65, F.R.G.

Kieran F. Geoghegan
Pfizer Central Research, Eastern Point Road, Groton, Connecticut 06340, U.S.A.

Stephan K. Grant
Smith Kline Beecham Pharamaceuticals, Research Laboratories, King of Prussia, Pennsylvania 19406, U.S.A.

Stephen Hawrylik
Pfizer Central Research, Eastern Point Road, Groton, Connecticut 06340, U.S.A.

William F. Heath, Jr.
Division of Biochemistry, Lilly Research Laboratories, Indianapolis, Indiana 46285, U.S.A.

Andrew Hemmings
Laboratory of Molecular Biology, Department of Crystallography, Birkbeck College, Malet Street, London WC1E 7HX, London, U.K.

Peter M. Hobart
Pfizer Central Research, Eastern Point Road, Groton, Connecticut 06340, U.S.A.

Stefan Höglund
Institute of Biochemistry, Biomedical Centre, S-751 23 Uppsala, Sweden.

Eric Hunter
Department of Microbiology, The University of Alabama at Birmingham, UAB Station, Birmingham, Alabama 35294, U.S.A.

Kazuyoshi Ikuta
Institute of Immunological Science, Hokkaido University, Sapporo, Japan.

Mariusz Jaskólski
Crystallographic Laboratory, NCI-Frederick Cancer Research Facility, BRI-Basic Research Program, P.O. Box b, Frederick MD 21701, U.S.A.

Ian M. Jones
N.E.R.C. Institute of Virology, Mansfield Road, Oxford, OX1 3SR, U.K.

U. Junker
Max von Pettenkofer Institut, University of Munich, Pettenkoferstr. 9a, D-8000 Munich 2, F.R.G.

Iyoko Katoh
Cellular Technology Institute, Otsuka Pharamceutical Co. Ltd., Tokushima 771-01, Japan.

John Kay
Department of Biochemistry, University of Wales, P.O. Box 903, Cardiff CF1 1ST, Wales, U.K.

Stephen B.H. Kent
Graduate School of Science and Technology, Bond University, Queensland 4229, Australia

Alan J. Kingsman
Department of Biochemistry, University of Oxford, South Parks Road, Oxford OX1 3QU, U.K.

Susan M. Kingsman
Department of Biochemistry, University of Oxford, South Parks Road, Oxford OX1 3QU, U.K.

Marie-Therèse Knoop
Max Planck-Institut für Molekular Genetik, Abt. Schuster, Ihnstr. 73, D-1000 Berlin 33, F.R.G.

V. Kruft
Max Planck-Institut für Molekular Genetik, Ihnstr. 73, D-1000 Berlin 33, F.R.G.

Mei-Huei T. Lai
Division of Molecular Biology, Lilly Research Laboratories, Indianapolis, Indiana 46285, U.S.A.

Risto Lapatto
Laboratory of Molecular Biology, Department of Crystallography, Birkbeck College, Malet Street, London WC1E 7HX, London, U.K.

S. Edward Lee
Pfizer Central Research, Eastern Point Road, Groton, Connecticut 06340, U.S.A.

Ronald Luftig
Department of Microbiology, Immunology and Parasitology, Louisiana State University Medical Center, New Orleans, LA, U.S.A.

Garland R. Marshall
Department of Pharamacology, Washington University School of Medecine, St. Louis, MO 63110, U.S.A.

James Merson
Pfizer Central Research, Sandwich, Kent, U.K.

Maria Miller
Crystallographic Laboratory, NCI-Frederick Cancer Research Facility, BRI-Basic Research Program, P.O. Box b, Frederick MD 21701, U.S.A.

Helen R. Mills
N.E.R.C. Institute of Virology, Mansfield Road, Oxford, OX1 3SR, U.K.

Karin Moelling
Max Planck-Institut für Molekular Genetik, Ihnstr. 73, D-1000 Berlin 33, F.R.G.

Michael L. Moore
Smith Kline Beecham Pharamaceuticals, Research Laboratories, King of Prussia, Pennsylvania 19406, U.S.A.

M. Nawrath
Max Planck-Institut für Molekular Genetik, Ihnstr. 73, D-1000 Berlin 33, F.R.G.

Åsa Nilsson
Institute of Biochemistry, Biomedical Centre, S-751 23 Uppsala, Sweden.

Kohei Oda
Laboratory of Applied Microbiology, Department of Agricultural Chemistry, University of Osaka Prefecture, Sakai, Osaka 591, Japan.

Lars-Göran Öfverstedt
Institute of Biochemistry, Biomedical Centre, S-751 23 Uppsala, Sweden.

Stephen Oroszlan
Laboratory of Molecular Virology and Carcinogenesis, BRI-Basic Research Program, NCI-Frederick Cancer Research Facility, Frederick, MD 21701, U.S.A.

John Overington
Laboratory of Molecular Biology, Department of Crystallography, Birkbeck College, Malet Street, London WC1E 7HX, London, U.K.

Hilary A. Overton
N.E.R.C. Institute of Virology, Mansfield Road, Oxford, OX1 3SR, U.K.

Muhsin Özel
Institute of Virology, Robert Koch-Institute of the Federal Health Administration, D-1000 Berlin 65, F.R.G.

L. Pavlitzkova
Institute of Organic Chemistry and Biochemistry, Czechoslovak academy of Science, Flemingovo 2, 16610 Prague, Czechoslovakia.

Laurence H. Pearl
Department of Biochemistry, University College London, Gower Street, London WC1E 6BT, U.K.

Cheng Peng
Division of Molecular Virology, Baylor College of Medicine, One Baylor Plaza, Houston, Texas 77030, U.S.A.

Jim Pitts
Laboratory of Molecular Biology, Department of Crystallography, Birkbeck College, Malet Street, London WC1E 7HX, London, U.K.

Sung Rhee
Department of Microbiology, The University of Alabama at Birmingham, UAB Station, Birmingham, Alabama 35294, U.S.A.

Daniel Rich
School of Pharmacy, The University of Wisconsin, 425 North Charter Street, Madison, Wisconsin 53706, U.S.A.

M.M. Roberts
Laboratory of Molecular Virology and Carcinogenesis, BRI-Basic Research Program, NCI-Frederick Cancer Research Facility, Frederick, MD 21701, U.S.A.

E.M.J. Roud Mayne
Department of Genetics, Glaxo Group Research Ltd., Greenford, U.K.

Bangalore K. Sathyanarayana
Crystallographic Laboratory, NCI-Frederick Cancer Research Facility, BRI-Basic Research Program, P.O. Box b, Frederick MD 21701, U.S.A.

Maurice E. Scheetz
Division of Immunology, Lilly Research Laboratories, Indianapolis, Indiana 46285, U.S.A.

Kathryn Scheld
Pfizer Central Research, Eastern Point Road, Groton, Connecticut 06340, U.S.A.

Thomas Schulze
Max Planck-Institut für Molekular Genetik, Ihnstr. 73, D-1000 Berlin 33, F.R.G.

S. Seelmeir
Max von Pettenkofer Institut, University of Munich, Pettenkoferstr. 9a, D-8000 Munich 2, F.R.G.

Oncar M.P. Singh
Department of Genetics, Glaxo Group Research Ltd., Greenford, U.K.

Ulf Skoglund
Department of Molecular Genetics, Medical Nobel Institute, Karolinska Institute, Box 604000, S-104 01 Stockholm, Sweden.

M. Soucek
Institute of Organic Chemistry and Biochemistry, Czechoslovak Academy of Science, Flemingovo 2, 16610 Prague, Czechoslovakia.

Amy L. Swain
Crystallographic Laboratory, NCI-Frederick Cancer Research Facility, BRI-Basic Research Program, P.O. Box B, Frederick MD 21701, U.S.A.

Klaus von der Helm
Max von Pettenkofer Institut, University of Munich, Pettenkoferstr. 9a, D-8000 Munich 2, F.R.G.

M.P. Weir
Department of Genetics, Glaxo Group Research Ltd., Greenford, U.K.

Peter Whittle
Pfizer Central Research, Sandwich, Kent, U.K.

Andrew Wilderspin
Laboratory of Molecular Biology, Department of Crystallography, Birkbeck College, Malet Street, London WC1E 7HX, London, U.K.

Wilma Wilson
Department of Biochemistry, University of Oxford, South Parks Road, Oxford OX1 3QU, U.K.

Thorsten Winkel
Institute of Virology, Robert Koch-Institute of the Federal Health Administration, D-1000 Berlin 65, F.R.G.

Alexander Wlodawer
Crystallographic Laboratory, NCI-Frederick Cancer Research Facility, BRI-Basic Research Program, P.O. Box b, Frederick MD 21701, U.S.A.

Stephen Wood
Laboratory of Molecular Biology, Department of Crystallography, Birkbeck College, Malet Street, London WC1E 7HX, London, U.K.

Yoshiyuki Yoshinaka
Tsukuba Life Science Center, The Institute of Physical and Chemical Research, Tsukuba, Ibaraki 305, Japan

Peter H. Zervos
Division of Molecular Biology Lilly Research Laboratories, Indianapolis, Indiana 46285, U.S.A.

Zhang-Yang Zhu
Laboratory of Molecular Biology, Department of Crystallography, Birkbeck College, Malet Street, London WC1E 7HX, London, U.K.

Introduction

Laurence H. Pearl

For several years since the original observations (von der Helm, 1977; Yoshinaka and Luftig, 1977) that retroviruses encoded a proteolytic activity of their own, the retroviral proteases were rather enigmatic proteins, pushed into the background by their glamourous retroviral siblings, reverse transcriptase and the viral oncogenes. Nonetheless, the continuing interest and careful work of several key laboratories, progressively assembled a body of genetic and biochemical data which increasingly pointed to a very fundamental role for the protease in a retrovirus.

In the early eighties, two events sparked a major resurgence of interest in retroviral proteases; the first of these was the observation by Hiroyuki Toh and colleagues (Toh *et al*, 1985), of homology between a conserved triplet of amino acids (Asp-Thr/Ser-Gly) in retroviral proteases, and the active site sequence of a known class of enzymes, the aspartic proteinases. The second event, was the realisation that a new infectious and lethal disease, the acquired immunodeficiency syndrome, was caused by a replication-competant retrovirus.

Following Toh's observations, retroviral proteases became the subject of considerable structural investigation, both by theoretical and modelling methods, and by X-ray crystallography. At the time of writing, four independant crystal structure determinations of retroviral proteases (Rous sarcoma virus once; HIV-1 three times) have been published, and have confirmed the active site homology observed by Toh, and the necessity of dimerisation as we had predicted (Pearl and Taylor, 1987). Site directed mutagenesis of the protease active site aspartate in an otherwise competant provirus (Kohl *et al*, 1988), showed that protease function was essential to viral replication and that the protease was therefore an important target for drug development. Fortunately, the very close mechanistic relationship to the well described mammalian and fungal aspartic proteinases, has enabled the rapid transfer of a large body of medicinal chemistry expertise, acquired in the development of renin inhibitors, into the production of highly active and specific inhibitors of the HIV proteinase, which are now the major leads in the development of a safe and effective chemotherapy for AIDS.

The role of the protease is the hydrolysis of several very specific peptide bonds in the *gag* and *gag-pol* polyproteins which are the major primary translational products of the unspliced viral mRNA, to produce the mature *gag* and *pol* encoded proteins. It is this activity that produces the transition to the mature infectious form of the virus. Paradoxically, the protease itself is one of the protein domains in the *gag-pol* polyprotein and is responsible for its own excision from this precursor form. The ability of the protease to cut itself out of the *gag-pol* polyprotein is in some way dependant on the assembly of the *gag* and *gag-pol* precursors, which is in its turn a function of the carefully controlled stoichiometry of these two polyproteins (Felsenstein and Goff, 1988). The protease is emerging as the lynch-pin in the whole process of morphogenesis and maturation, so critical to the virus life-cycle.

This book arises from a meeting sponsored by the U.K. Medical Research Council AIDS Directed Programme, and held at the Royal Hotel, York, England, on the 4th and 5th of December 1989.

This meeting was an attempt to bring together a group of key researchers, from very different backgrounds, but sharing a common interest in understanding the biological mechanisms by which assembly and maturation of retroviruses takes place, and the central role in these processes, played by the viral protease. The expertise of the speakers at the meeting, ranged over the entire gamut of modern virology, molecular biology, biochemistry and biophysics, and provided an exciting multidisciplinary atmosphere, resulting in many new ideas and collaborations. We hope that this book will transmit that excitement to the reader and act as a stimulus to research in this field.

ACKNOWLEDGEMENTS

I am extremely grateful to Mr. David Grist of Macmillans for his enthusiasm for this book, and to the authors for their great speed in providing their manuscripts, and for (mainly) following the guidelines on presentation. An editor's job is never a simple one, but they have made mine simpler than most.

I gratefully acknowledge the Medical Research Council AIDS Directed Programme, for their generous financial support for this meeting and for my own research in this area. I particularly wish to thank Dr. Geoffrey Schild (the Director of the AIDS Directed Programme), and Mr. Peter French, Dr. Angela Williams and Dr. Jane Cope of the Programme Secretariat, for their enthusiasm and considerable assistance with the organisation of this meeting. Thanks are also due to my colleagues at U.C.L. for their patience during the preparation of this book and to the Department of Biochemistry for provision of excellent document preparation facilities. Finally, special thanks are due to Professor Robin Weiss and his group at the Institute of Cancer Research, for their unwavering interest and support during academically trying times, and for teaching a crystallographer how to clone genes.

This book is dedicated to all those in the world for whom the research contained therein offers hope.

REFERENCES

Felsenstein, K.M. and Goff, S.P. *J. Virol.*, **62**, 2179–2184, (1988).
Kohl, N.E., Emini,E.A., Schleif,W.A., Davis,L.J., Heimbach,J.C., Dixon,R.A.F., Scolnick,E.M. and Sigal,I.S. *Proc. Natl. Acad. Sci USA*, **85**, 4686–4690, (1988).
Pearl, L.H. and Taylor, W.R. *Nature*, **329**, 351–354, (1987).
Toh, H., Onon, M., Saigo, K. and Miyata, T. *Nature*, **315**, 691, (1985).
von der Helm, K. *Proc. Natl. Acad. Sci. USA*, **74**, 911–915, (1977).
Yoshinaka, Y. and Luftig, R.B. *Cell*, **12**, 709–719, (1977).

1
Characterisation and Inhibition of the Retroviral HIV-Protease

Klaus von der Helm, S. Seelmeir and U. Junker

Virus encoded proteases were first described in avian and murine retroviruses (von der Helm, 1977; Yoshinaka and Luftig, 1977). They are 'processing' proteases cleaving virus encoded polyproteins into smaller proteins which then function as viral structural (scaffolding) proteins or virus specific enzymes. Thus, they are anabolic rather than catabolic enzymes. They have been demonstrated in many viruses different from retroviruses (for review see Kräusslich and Wimmer, 1988). In retroviruses they were shown to process the *gag-pol* protein precursor to yield proteins required for production of infectious virus. It has been shown that replication of murine retrovirus with a defective protease yielded immature virus particles lacking infectivity (Crawford and Goff, 1985; Katoh *et al.*, 1985). Replication of HIV (human immunodeficiency virus) carrying a lethal protease mutant led to non-infectious virus particles (Kohl, *et al.*, 1988). Retroviral proteases appear to have a highly specific relation to their virus encoded substrate (Dittmar and Moelling, 1978; Khan and Stephenson, 1979; Yoshinaka *et al.*, 1985; Kräusslich and von der Helm, 1987). Examination of the amino acid sequences of retroviral proteases indicated the presence of a conserved Asp-Thr(Ser)-Gly sequence, suggesting that they are aspartic-type proteases (Kräusslich and Wimmer, 1988; Toh *et al.*, 1985; Pearl and Taylor, 1987). By site specific mutagenesis of the essential Asp-25 of the HIV protease (PR) performed by recombinant technique we and others could prove directly that the PR has an aspartic proteolytic site (Kohl *et al.*, 1988; Mous *et al.*, 1988; Seelmeir *et al.*, 1988; Loeb *et al.*, 1989) because the Asp-25 mutation rendered the enzyme inactive. The enzyme is proteolytically active only as a (homo)-dimer of two of the genetically encoded monomers (Meek *et al.*, 1989; Katoh *et al.*, 1989) as theoretically predicted (Pearl and Taylor, 1987).

RESULTS AND DISCUSSION

Proteolytic activity

As protease is an internal part of the *gag-pol* precursor protein the question arose of how PR is cleaved from the precursor and activated.

This question, too, was answered by using the Asp-25 mutant (Mous *et al.*, 1988; Seelmeir *et al.*, 1988; Loeb *et al.*, 1989). The active site aspartate of the HIV-protease is also responsible for the enzyme activation because the mutant protease is not cleaved off. It is not clear whether an intermolecular or intramolecular autocatalytic mechanism is involved in the PR activation.

Using a further mutant, Asp60->Thr, which mutation maps distantly from the active center of the HIV PR and is obviously not part of it we could demonstrate that this PR mutant had a significantly reduced level of self-processing and activation (Junker *et al.*,unpublished). Also, the RT protein 51 kd was processed very inefficiently from the PR/D60-RT precursor. We suggest that this mutation puts a confirmational strain into the PR-RT polyprotein thus hindering the autocatalytic cleavage and activation of PR and processing of the RT 51kd. Once the D-60 mutant enzyme has been cleaved off by means of resin-immobilized active WT-protease it appeared to function normally.

The topology of activation and proteolytic action of the PR is not precisely determined. It is conceivable that the enzyme processes the polyprotein at or near the cytoplasmic membrane during particle budding or that the proteolytic process occurs extracellularly within budded but not matured particles. Former studies on the topology of processing/ assembly of MLV (Witte and Baltimore, 1978) suggested the membrane as the maturation site.

The results of a preliminary experiment seem to favour the first possibility: We harvested HIV from infected H9 cell culture medium at very short and long time intervals, i.e. at 20 min or 24 hrs (von der Helm *et al.*, 1989a). After purifying both virus populations we compared their degree of *gag* precursor processing by Western blot analysis with anti-*gag* serum. Using a rapidly replicating HIV isolate (Guertler, unpublished) we found a very similar processing pattern in both, differently aged virus populations: practically all *gag* precursor had been matured to the individual *gag* proteins. This suggests that at least in this rapidly replicating HIV isolate no significant maturation of virus capsid occurred extracellularly unless this would have taken place with in the first 20 minutes after budding. In a more slowly replicating HIV isolate more of the unprocessed gag precursor was detectable similarly in both, the 20 min and 24 hrs population. This again indicated that despite insufficient intracellular processing no further extracellular processing had taken place.

If *gag* precursor is not or not sufficiently cleaved intracellularly (as seen in the latter experiment), it is nevertheless released from the cell (Kohl *et al.*, 1988; Göttlinger *et al.*, 1989). Uncleaved *gag* polyprotein is also released from cells infected with competent vaccinia virus containing recombinantly inserted HIV-*gag* (with out PR) sequences. We measured the density and the Svedberg-constant *S* of this released material by means of respective sucrose gradient centrifugations as described before for RSV (Canaani *et al.*, 1973). The buoying density of the uncleaved recombinant HIV gag precursor material was about 1.8-1.9. This is a slightly higher density than that of retroviruses (1.6-1.75) and this result indicates that the *gag* material contains probably a slightly larger amount of nucleic acid (presumably RNA) than retrovirions. When we analyzed the material by velocity gradient centrifugation we found a definite 'smearing' through the gradient, i.e. *S*-values of less and much more than 600 which is the Svedberg-constant for RSV (Canaani *et al.*), suggesting various sizes of (maybe ordered) *gag*-RNA-polyprotein aggregates.

Even non-matured retroviruses have a Svedberg-constant and size comparable to matured virions. Thus, these results argue that at least a minimal part of a 'trimming'-maturation process is going on even before budding of immature virus.

Inhibition

As the PR is an aspartic protease we and others tested in an *in vitro* system of recombinant viral protease and gag precursor the inhibiting effect of Pepstatin A and found the proteolytic the proteolytic activity was inhibited at a concentration of about 10^{-4} M (IC50). As Pepstatin A is hydrophobic and possibly not well cell-permeable we checked directly to see whether or not Pepstatin A would inhibit the in vivo HIV-*gag* processing in infected cell culture and reduce the level of infectious HIV. Thus we measured the HIV antigen (CA) production, the formation of active reverse transcriptase (RT) and the infectivity of HIV still released into the culture medium (von der Helm *et al.*, 1989b). After incubation of HIV-infected H9 cells with Pepstatin A for 2, 4, or 11 days the culture medium contained significantly less HIV *gag*-antigen (determined by anti CA(p24) serum) and no or borderline RT activity. Moreover, after Pepstatin A treatment for 2 or 4 days, no more infectious HIV could be detected in the culture medium.

These results are interesting because in earlier cell culture experiments with MLV containing a defect protease (Crawford and Goff, 1985; Katoh *et al.*, 1985) or with HIV containing an inactive PR Asp-25 mutant (Kohl *et al.*, 1988) a significantly larger amount of (non-infectious) viral antigen was produced than in our *in vivo* inhibition by Pepstatin A. The reason for this is not known; it might be that Pepstatin A not only inhibits the virus protease activity but also affects other steps in the HIV assembly/maturation. These considerations are speculative since we do not know where the PR and consequently where pepstatin A acts. Does it (i) penetrate the cytoplasmic membrane and act within the membrane, or (ii) does it act only extracellularly on released but immature virus, preventing its maturation?

From results discussed above we would favour the first possibility as the more likely one. This does however not exclude an additional extracellular processing by the viral PR.

ACKNOWLEDGEMENTS

We thank H.Wolf for the vaccinia-*gag* recombinant clone and the Bundesministerium für Forschung und Technologie for financial support.

REFERENCES

Canaani,E., von der Helm,K., and Duesberg,P. *Proc.Natl.Acad.Sci.USA*, **70**, 401-405, (1973).

Crawford,S. and Goff,S.P. *J.Virol.* **53**, 899-907, (1985).

Dittmar,K.J. and Moelling,K. *J.Virol.*, **28**, 106-118, (1978).

Göttlinger,H.G., Sodroski,J.G., Haseltine,W.A. *Proc.Natl.Acad.Sci.USA*, **86**, 5781-5785, (1989).

Katoh,I., Yoshinaka,Y.,, Rein,A., Shibuya,M., Odaka,T. and Oroszlan,S. *Virology*, **145**, 280-292, (1985).

Katoh,I., Ikawa,Y., and Yoshinaka,Y. *J.Virol.*, **63**, 2226-2232, (1989).

Khan,A.S. and Stephenson,J.R. *J.Virol.*, **29**, 649-656, (1979).

Kohl, N.E., Emini,E.A., Schleif,W.A., Davis,L.J., Heimbach,J.C., Dixon,R.A.F., Scolnick,E.M. and Sigal,I.S. *Proc. Natl. Acad. Sci USA*, **85**,'4686-4690, (1988).

Kräusslich,H.G. and von der Helm,K. *Virology*, **156**, 246-252, (1987).

Kräusslich,H.G. and Wimmer,E. *Ann.Rev.Biochem.*, **57**, 701-754, (1988).

Loeb,D.D., Hutchison III,C.A., Egdell,M.H., Farmerie,W.G. and Swanstrom,R. *J. Virol.*, **63**, 111-119, (1989).

Meek,T.D., Dayton,B.D., Metcalf,B.W., Dreyer,G.B., Strickler,J.E.,Gorniak,J.G., Rosenberg,M., Moore,M.L., Magaard,V.W., Debouck,C. *Proc.Natl.Acad.Sci.USA*, **86**, 1841-1845, (1989).

Mous,J., Heimer,E.P., and LeGrice,St.F.J. *J.Virol.*, **62**, 1433-1436, (1988).

Pearl,L.H. and Taylor,W.R. *Nature*, **329**, 351-354, (1987).

Scolnick,E.M. and Sigal,I.S. *Proc.Natl.Acad.Sci.USA*, **85**, 4686-4690, (1988).

Seelmeir,S., Schmidt,H., Turk,V. and von der Helm,K. *Proc.Natl.Acad.Sci.USA*, **85**, 6612-6616, (1988).

Toh,H., Ono,M., Saigo,K. and Miyata T. *Nature*, **315**, 691-692, (1985).

von der Helm,K. *Proc.Natl.Acad.Sci.USA*, **74**, 911-915, (1977).

von der Helm,K., Junker,U., Weiss,S., Reinbold,G. and Seelmeir,S. in 'Viral Proteinases as Targets of Chemotherapy (H-G.Kräusslich, S.Oroszlan and E.Wimmer, eds.) Cold Spring Harbor Laboratory Press p.141, (1989a).

Von der Helm.K., Gürtler,L., Eberle,J. and Deinhardt,F. *FEBS Lett.*, **247**, 349-352, (1989b).

Witte,O. and Baltimore,D. *J. Virol.*, **26**, 750-756, (1978).

Yoshinaka,Y. and Luftig,R.B. *Cell*, **12**, 709-719, (1977).

Yoshinaka,Y., Katoh,I., Copeland,T.D. and Oroszlan,S. *J.Virol.*, **55**, 870-873, (1985).

2
The HIV-1 Aspartyl Protease: Maturation and Substrate Specificity

Christine Debouck, Ingrid C. Deckman, Stephan K. Grant,
Robert J. Craig and Michael L. Moore

In all retroviruses, the viral aspartyl protease is first translated as part of a large polyprotein precursor, the 160 kilodalton *gag-pol* polyprotein in the case of HIV-1 (Rey *et al*, 1987; Jacks *et al*, 1988). Experimental evidence strongly suggests that HIV-1 viral particles initially bud out of the host cell with an immature structure and composition, containing unprocessed $Pr55^{gag}$ and $Pr160^{gag-pol}$ polyproteins (Hockley *et al*, 1988; Gottlinger *et al*, 1989, Peng *et al*, 1989). The question of when and how the protease actually processes these polyproteins is important for the understanding of the viral life cycle but remains to be answered. In an effort to gain information on this process, we undertook a series of genetic studies in *Escherichia coli* on the maturation of the protease from its precursor forms and on this protease substrate preferences. This information should not only contribute to our understanding of the viral maturation process, but also assist us in the design of inhibitors of this essential viral enzyme.

STRUCTURE AND ACTIVITY OF PROTEASE CONSTRUCTS

The mature, 99 amino acid-long HIV-1 protease has been expressed in bacteria by several laboratories using genetic constructs that encode the 99-mer species itself or, more typically, a precursor species that is readily processed to the mature form of the enzyme (see 'Viral Proteinases as Targets for Chemotherapy' (1989) as a source of information and references). As a rule, protease with the best attributes of solubility, homogeneity and activity has been obtained from those expression vectors that direct the production of the mature, dimeric protease from a precursor form. Examples of protease constructs that were generated in our laboratory are shown in Figure 1 and are described hereafter.

In our laboratory, the protease expression vectors that yielded the mature HIV-1 protease in its most soluble, homogenous and active form were the ones encoding the PRO3 and PRO4 constructs. In these constructs, the protease was expressed via 'auto'-processing (probably an intermolecular event) of a larger precursor containing both upstream sequences (11 or 56 residues for PRO3 and PRO4, respectively) and downstream sequences (18 residues for both PRO3 and PRO4) from the *pol*

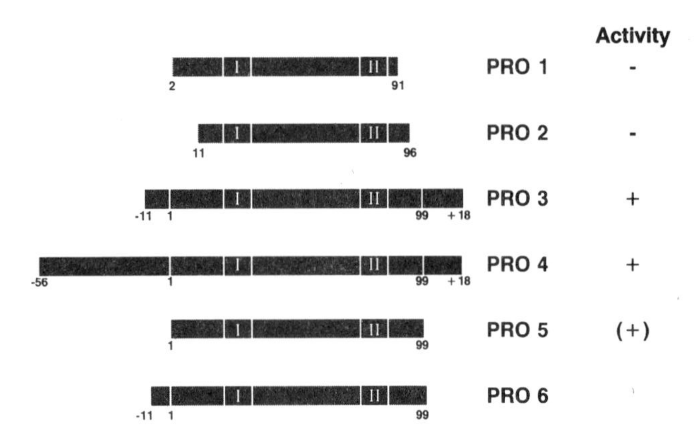

Figure 1
Schematic representation of HIV-1 protease genetic constructs. The HIV-1 protease coding region present in six bacterial expression vectors (PRO1 to PRO6) is depicted. PRO1-PRO4 have been previously described (Debouck *et al*, 1987). PRO5 and PRO6 are described in the text. The boxed regions marked I and II correspond to the two domains that are highly conserved among retroviral, microbial and cellular aspartyl proteases (Pearl and Taylor, 1987). The numbering starts with 1 at the first residue (proline) and 99 at the last residue (phenylalanine) of the mature HIV-1 protease;-11 and -56 indicate the number of amino acid residues from the *pol* region upstream of the mature protease; +18 indicates the number of residues from the *pol* region downstream of the mature protease. The activity of the protease products against $Pr55^{gag}$ in *E. coli* (co-expression system) is shown: +, active, Pr55 is processed; -, inactive, Pr55 is not processed.

open reading frame. High levels of protease activity were readily demonstrated for the PRO3 and PRO4 constructs both within whole bacterial cells using a co-expressed $Pr55^{gag}$ polyprotein substrate (Debouck *et al*, 1987) and *in vitro* after purification of the enzyme (Meek *et al*, 1989; Strickler *et al*, 1989). We also constructed an expression vector, PRO5, for the direct production of the mature 99 amino acid protease. For this purpose, a translation initiation codon (ATG) was placed immediately before the first codon of the protease (encoding proline 1) and a stop codon (TAG) immediately following the last codon (encoding phenylalanine 99). This protease construct was found to be active when tested in whole bacterial cells for the specific processing of a co-expressed $Pr55^{gag}$ substrate. However, PRO5 yielded a mature protease that accumulated as insoluble inclusion bodies that required solubilization by denaturing agents and subsequent refolding prior to the detection of enzymatic activity *in vitro*. Constructs designed for the expression of slightly truncated forms of the protease (i.e. PRO1 and PRO2) were found to be completely inactive (Debouck *et al*, 1987), indicating that amino acid residues that constitute the amino- and carboxyl-termini of the enzyme are critical for its structure and/or function. Recent information on the crystal structure of the HIV-1 protease has revealed that these residues actually form the interface between the subunits of this homodimeric enzyme (Lapatto *et al*, 1989; Wlodawer *et al*, 1989).

When produced from the PRO3 and PRO4 expression vectors, the protease precursors do not accumulate in detectable amounts but undergo instead a 'self'-processing (most probably by intermolecular catalysis (Wlodawer *et al*, 1989)) to yield the mature 99-mer protease (Fig.2). Using an *E. coli* host strain deficient in the heat-shock response, we found that the product of PRO4 accumulated in its precursor form as insoluble inclusion bodies that could readily be solubilized with urea and dithiothreitol (Strickler *et al*, 1989). Upon refolding, the purified precursor species rapidly converted to the mature 99-mer protease species as observed by Western blot analysis. It was observed, however, that cleavage at the Phe-Pro bond constituting the carboxyl-terminus of the mature protease took place significantly slower than cleavage at its amino-terminal Phe-Pro bond (Strickler *et al*, 1989). We recently constructed an expression vector, PRO6, which is identical to PRO3 except for a translation stop codon placed immediately after the last residue of the mature protease (i.e. phenylalanine 99). Preliminary results revealed that, in PRO6, the proteolytic cleavage at the remaining amino-terminal Phe-Pro bond is very slow compared to the processing at that site in PRO3. Similar results have been reported by Hostomsky *et al* (1989). This suggests that the presence of additional, cleavable amino acid residues before the amino-terminus and after the carboxyl-terminus of the mature protease is important for the proper folding and maturation of the enzyme. Deletion of one (as in PRO6) or both (as in PRO5) precursor regions clearly results in improper folding and concomitant poor 'self'-processing (PRO6) or poor solubility (PRO5). Obviously, these results have all been obtained in bacteria, but it is likely that the protease precursor also plays a critical role in the proper folding of the protease domain within the *gag-pol* precursor in virally infected cells.

In an effort to examine the proteolytic activity of the protease precursor forms towards the $Pr55^{gag}$ polyprotein, we recently constructed derivatives of PRO4 in which one or both of the Phe-Pro bonds that bracket the mature protease have been altered by site-directed mutagenesis (Fig.2). We chose initially to change each site by severe mutation without altering the primary sequence of the 99-mer protease itself. Phenylalanine -1 was changed to aspartic acid, proline 100 was changed to aspartic acid, and both mutations were combined in one construct. When these constructs were co-expressed with $Pr55^{gag}$ in *E. coli*, we found that both PRO4(F-1D) and PRO4(P100D) efficiently processed $Pr55^{gag}$, although some alternate processing pattern was observed with F-1D, probably the result of altered proteolytic specificity. Interestingly, the double mutant also exhibited $Pr55^{gag}$ processing, but it was dramatically reduced. It is possible that the reduced proteolytic activity of the double mutant is simply due to the presence of undesirable negative charges within these hydrophobic regions of the protein. Alternatively, the much decreased activity of this double mutant raises the possibility that the protease domain within the *gag-pol* precursor is itself poorly active, reminiscent of the zymogen form of other proteases. Experiments are in progress to introduce other mutations at these Phe-Pro bonds and to examine their proteolytic activity.

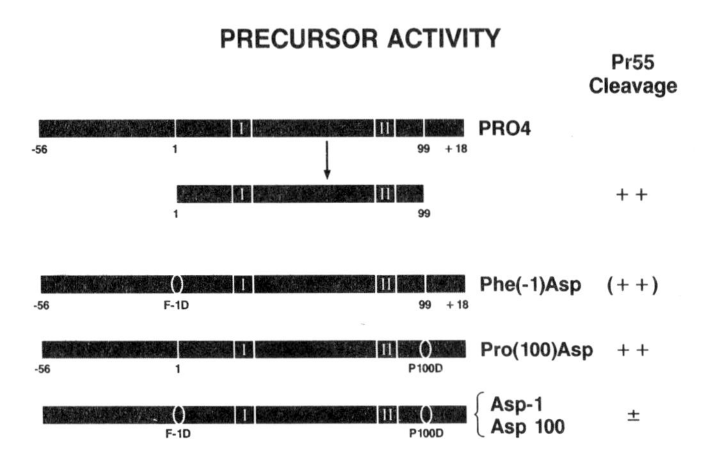

Figure 2
Structure-activity of HIV-1 protease precursors expressed in bacteria. The PRO4 construct is schematically represented using the same symbols as in Figure 1. Three mutated forms of PRO4 are also shown: in one mutant (F-1D), phenylalanine -1 was changed to aspartic acid by site-directed mutagenesis; in the other mutant (P100D), proline 100 was changed to aspartic acid; the last mutant (F-1D;P100D) comprises both mutations. The activity of these protease precursor forms against $Pr55^{gag}$ in *E. coli* (co-expression system) is shown: ++, active, 100% processing of $Pr55^{gag}$; +-, poorly active, <10% processing of $Pr55^{gag}$; (++), indicates alternate processing pattern.

SUBSTRATE SPECIFICITY OF THE HIV-1 PROTEASE

Seven sites within the HIV-1 *gag-pol* polyprotein have been known to correspond to cleavage sites for the HIV-1 protease (sites 1 to 6 and 8, Fig.3, Henderson *et al*, 1988). In addition, site 7 (Fig.3) located within the reverse transcriptase coding sequence has been shown to be an HIV-1 protease cleavage site in our laboratory (Mizrahi *et al*, 1989). Indeed, using our double-plasmid expression system, we co-expressed in *E. coli* a precursor form of reverse transcriptase together with the HIV-1 protease and showed that the carboxyl- terminus of the 51-kilodalton chain of reverse transcriptase is the result of an HIV-1 proteolytic cleavage at a Phe-Tyr bond. It remains to be determined, however, if this site is recognized in virally infected cells (the carboxyl- terminus of p51 reverse transcriptase from viral particles has not been sequenced to date).

SITES IN HIV-1 GAG-POL

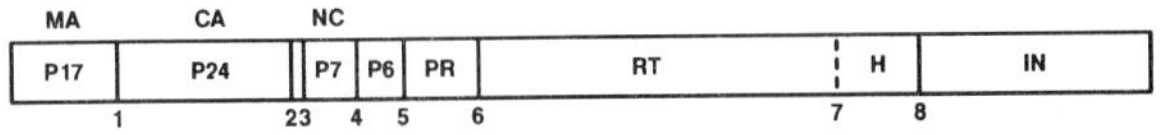

MA CA NC

P17 | P24 | P7 | P6 | PR | RT | H | IN

1 23 4 5 6 7 8

1	P17 * P24	SER • GLN • ASN • TYR * PRO • ILE • VAL • GLN
2	P24 * X	ALA • ARG • VAL • LEU * ALA • GLU • ALA • MET
3	X * P7	ALA • THR • ILE • MET * MET • GLN • ARG • GLY
4	P7 * P6	PRO • GLY • ASN • PHE * LEU • GLN • SER • ARG
5	* PR	SER • PHE • ASN • PHE * PRO • GLN • ILE • THR
6	PR * RT	THR • LEU • ASN • PHE * PRO • ILE • SER • PRO
7	RT51 * RNase H	ALA • GLU • THR • PHE * TYR • VAL • ASP • GLY
8	RT * IN	ARG • LYS • ILE • LEU * PHE • LEU • ASP • GLY

Figure 3
Cleavage sites for the HIV-1 protease within the HIV-1 *gag-pol* precursor. The Pr160$^{gag-pol}$ precursor is represented schematically. It comprises the p17 matrix protein (MA), the p24 capsid protein (CA), the p7 nucleocapsid protein (NC), the p6 protein, the protease (PR), the reverse transcriptase (RT) with its carboxyl-terminal ribonuclease H domain (H) and the endonuclease or integrase (IN). The positions and primary sequences of eight known cleavage sites for the HIV-1 protease are indicated (Henderson *et al*, 1988; Mizrahi *et al*, 1989). The * represents the peptide bond that is cleaved by the HIV-1 protease. Amino acids are named using the conventional three-letter designation.

These cleavage sites do not exhibit major primary sequence identity at the exception of sites 1, 5 and 6 which have similar P4, P1 and P1' positions and match the consensus sequence Ser/Thr.Xaa.Yaa.Phe/Tyr*Pro (with cleavage at *). However, all eight sites have an overall hydrophobic composition. Residues in the P1 and P1' positions are strictly hydrophobic. Residues in P2 and P2' consist of hydrophobic or uncharged polar residues (with one exception in site 2). Charged residues are scarce in the P4 to P4' positions and are strictly (with one exception) excluded from the P2 to P2' positions. In order to determine whether or not the HIV-1 protease exhibits cleavage preference towards some sites versus the others, several laboratories have synthesized small peptides corresponding to the cleavage sites shown in Fig.3 (Billich *et al*, 1988; Darke *et al*, 1988; Schneider and Kent, 1988; Moore *et al*, 1989). Unfortunately, results from peptidolysis experiments have often been difficult to interpret because of the inherent hydrophobicity and subsequent poor solubility of these peptides. In an attempt to circumvent these technical difficulties, we chose to engineer each cleavage site individually within a heterologous protein. For this purpose, we selected galactokinase, the product of an *E. coli* gene which possesses convenient indicator properties (a galK^{+} phenotype results in red colonies on MacConkey indicator plates, whereas a galK^{-} phenotype gives white colonies). We initially inserted the p17-p24 cleavage site (site 1) in the middle of the galactokinase (galK) gene, away from the presumed active sites of this enzyme (Fig.4). Upon co-expression of this engineered galactokinase together with the HIV-1 protease in our double- plasmid expression system, we observed that galactokinase was completely processed in its two expected halves, indicating that it was a competent substrate for the HIV-1 protease in this system. Interestingly, this processing also resulted in the enzymatic inactivation of galactokinase, thereby providing us with a convenient colorimetric plate assay for HIV-1 protease activity. Indeed, in the absence of protease or in the presence of inactive protease, the bacterial colonies are red on appropriate indicator plates, whereas in the presence of active HIV-1 protease, the colonies are white on the same plates. We are now introducing all the other *gag-pol* cleavage sites at the same position within galactokinase in order to compare and hopefully quantify the HIV-1 proteolytic activity towards these various sites.

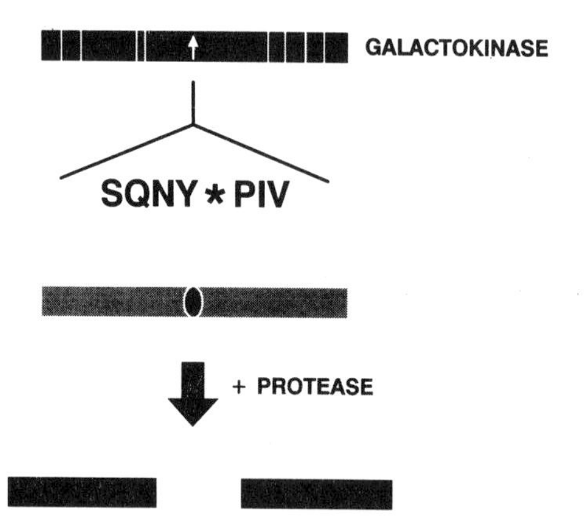

Figure 4
Engineering a heterologous HIV-1 protease substrate. The *E. coli* galactokinase gene is represented schematically with the regions conserved among galactokinase from different microorganisms highlighted. The site of in-frame insertion of an oligonucleotide encoding the HIV-1 protease cleavage site between p17 and p24 (site 1) is indicated by the arrow. Upon exposure to active HIV-1 protease, the engineered galactokinase is processed into the two expected halves as described in the text. The amino acid sequence of site 1 is shown using the conventional one-letter designation.

EXPRESSION AND ACTIVITY OF SIV PROTEASE

The animal retrovirus that is most closely related to HIV-1 is the simian immunodeficiency virus, SIV. The SIV protease coding sequence is very similar to the HIV-1 protease (48% amino acid sequence identity, 71% amino acid sequence similarity). Still, the SIV protease is sufficiently different from the HIV-1 enzyme to justify studies of its activity and substrate specificity as compared to the ones of its HIV-1 counterpart.

To this end, we have expressed the SIV protease in bacteria, using the same approach (Debouck *et al*, 1987) as for the expression of the HIV-1 protease. Briefly, a restriction endonuclease fragment encompassing some 270 codons from the beginning of the *pol* open reading frame was isolated from SIV_{MAC} proviral DNA. It was then inserted in our bacterial expression vector yielding a construct very similar to the HIV-1 PRO4 construct. As in HIV-1 PRO4, a 'self'-processed, mature 11-kilodalton protease species was produced upon induction and the precursor form did not accumulate to any detectable level (I. Deckman *et al*, manuscript in preparation). The protease thus produced cross-reacted with a rabbit polyclonal antibody raised against the HIV-1 protease, confirming the strong structural similarity of these two enzymes.

PEPTIDE SUBSTRATE CLEAVAGE

Cleavage site	Peptide	Km (mM) SIV_{MAC}	Km (mM) HIV-1
17 ↓ 24	Ac-SQNY ↓ PVV-NH_2	8.7	5.5
17 ↓ 24	Ac-RASQNY ↓ PVV-NH_2	3.5	5.5
	Ac-RASQNF ↓ PVV-NH_2	13	6.5
PR ↓ RT	Ac-ATLNF ↓ PISPIE-NH_2	7.0	6.7
RT51 ↓ 15	Ac-AETF ↓ YVD-NH_2	>10	4.4
RT ↓ ENDO	Ac-RKIL ↓ FLDG-NH_2	0.8	0.8

Figure 5

Comparative peptidolysis by recombinant SIV and HIV-1 proteases. The peptidolytic activity of highly purified SIV_{MAC} and HIV-1 proteases is presented using the Km kinetic constant obtained from standard Dixon plots. The peptide sequences corresponding to site 1 (p17*p24), site 6 (PR*RT), site 7 (RT51*15) and site 8 (RT*ENDO) are shown with the conventional amino acid one-letter designation. Certain amino acids were substituted in these sequences in order to facilitate synthesis or to improve solubility.

The recombinant SIV protease was also found to efficiently and specifically process the HIV-1 $Pr55^{gag}$ precursor when co-expressed in bacterial cells (Debouck *et al*, 1990). In order to better characterize the substrate specificity of the SIV protease and compare it to its HIV-1 congener, we purified the recombinant SIV protease and performed a series of peptidolytic assays using peptides corresponding to four HIV-1 *gag-pol* cleavage sites (Fig.5). All in all, the SIV protease exhibited similar if not identical lytic activity towards these peptides as compared to the purified HIV-1 protease, with the possible exception of cleavage at site 7. This site, a Phe-Tyr bond located between the p51 chain of reverse transcriptase and its ribonuclease H domain, appeared to be cleaved less efficiently by the SIV protease than by the HIV-1 protease. Since several peptides, corresponding to the other *gag-pol* cleavage sites, could not be tested (insolubility problem), we are now in the process of examining the SIV proteolytic activity towards the engineered galactokinase substrate containing each of the eight HIV-1 *gag-pol* cleavage sites. These studies should provide information on how primary sequence relates to the structure and function of these two highly homologous enzymes. Finally, analysis of the sensitivity of this and other animal retroviral proteases to inhibitors designed to block the HIV-1 protease will also yield information on structure-function relationships, in addition to repesenting an attractive therapeutic approach to the treatment of relevant animal retroviral infections.

ACKNOWLEDGMENTS

This work was supported in part by grants AI24845 and GM39526 from the National Institutes of Health.

REFERENCES

Billich,S., Knoop,M-T., Hansen,J., Strop,P., Sedlacek,J., Mertz,R. and Moelling,K. *J. Biol. Chem.*, **263**, 17905-17908, (1988).

Darke,P., Nutt,R.F., Brady,S.F., Garsky,V.M., Ciccarone,T.M., Leu,C-T., Lumma,P.K., Freidinger,R.M., Veber,D.F. and Sigal,I.S. *Bioch. Bioph. Res. Comm.* **156**, 297-303, (1988).

Debouck,C., Gorniak,J.G., Strickler,J.E., Meek,T.D., Metcalf,B.W. and Rosenberg,M. *Proc. Natl. Acad. Sci. USA.* **84**, 8903-8906, (1987).

Debouck,C., Deckman,I.C., Lazarus,G.M. and Mizrahi,V. in 'Human Retroviruses', (Groopman,J.E., Chen,I. Essex,M. and Weiss,R. eds.), A.R. Liss, New York, (1990).

Gottlinger,H.G., Sodroski, J.G. and Haseltine,W.A. *Proc. Natl. Acad. Sci. USA*, **86**, 5781-5785, (1989).

Henderson,L.E., Copeland,T.D., Sowder,R.C., Schultz,A.M. and Oroszlan,S. in 'Human Retroviruses, Cancer, and AIDS: Approaches to Prevention and Therapy', (Bolognesi,D. ed.), pp. 135-147, A.R. Liss, New York, (1988).

Hockley,D.J., Wood,R.D., Jacobs,J.P. and Garrett,A.J. *J. Gen. Virol.* **69**, 2455-2469, (1988).

Hostomsky,Z., Appelt,K. and Ogden,R.C. *Bioch. Bioph. Res. Comm.* **161**, 1056- 1063, (1989).

Jacks,T., Power,M.D., Masiarz,F.R., Luciw,P.A., Barr,P.J. and Varmus,H.E. *Nature* **331**, 280-283, (1988).

Lapatto,R., Blundell,T., Hemmings,A., Overington,J., Wilderspin,A., Wood, S., Merson,J.R., Whittle,P.J. Danley,D.E., Geoghegan,K.F., Hawrylik,S.J., Lee,S.E., Scheld,K.G. and Hobart,P.M. *Nature* **342**, 299-302, (1989).

Meek,T.D., Dayton,B.D., Metcalf,B.W., Dreyer,G.B., Strickler,J.E., Gorniak,J.G., Rosenberg,M., Moore,M.L., Magaard,V.W. and Debouck,C. *Proc. Natl. Acad, Sci. USA.* **86**, 1841-1845, (1989).

Mizrahi,V., Lazarus,G.M., Miles,L.M., Meyers,C.A. and Debouck,C. *Arch. Bioch. Bioph.* **273**, 347-358, (1989).

Moore,M.L., Bryan,W.M., Fakhoury,S.A., Magaard,V.W., Huffman,W.F., Dayton, B.D., Meek,T.D., Hyland,L., Dreyer,G.B., Metcalf,B.W., Strickler,J.E., Gorniak, J.G. and Debouck,C. *Bioch. Bioph. Res. Comm.* **159**, 420-425, (1989).

Pearl,L.H. and Taylor,W.R. *Nature* **329**, 351-354, (1987).
Peng.C., Ho,B.K., Chang,T.W. and Chang,N.T. *J. Virol.* **63**, 2550-2556, (1989).
Rey,F., Barre-Sinoussi,F. and Chermann,J.C. *Ann. Inst. Pasteur* **138**, 161-168, (1987).
Schneider,J. and Kent,S.B.H. *Cell* **54**, 363-368, (1988).
Strickler,J.E., Gorniak,J., Dayton,B., Meek,T., Moore,M., Magaard,V., Malinowski,J. and Debouck,C. *Proteins* **6**, 139-154, (1989).
Wlodawer,A., Miller,M., Jaskolski,M., Sathyanarayana, Baldwyn,E., Weber, I.T., Selk,L.M., Clawson,L., Schneider,J. and Kent,S.B.H. *Science* **245**, 616- 621, (1989).
'Viral Proteinases as Targets for Chemotherapy',(Krausslich,H-G., Oroszlan,S. and Wimmer,E. eds.), Cold Spring Harbor Laboratory Press,(1989).

3
Cleavage of RT/RNase H by HIV-1 Protease and Analysis of Substrate Cleavage Sites *in vitro*

K. Moelling, M. Nawrath, T. Schulze, L. Pavlitzkova, M. Soucek, K.-H. Budt, L. H. Pearl, M.-T. Knoop, J. Kay and V. Kruft

The *pol*-gene of HIV-1 expressed in bacteria undergoes autocatalytic processing which results in the generation of the p66 reverse transcriptase (RT)/RN ase H, the p32 endonuclease and the p9 protease. The partially purified protease generates *in vitro* from the p66 molecule a p66/p51 heterodimer, typical of the normally observed RT/RNase H, and a p15 carboxyterminal fragment. A synthetic peptide AETF'YVD derived from the p51/p15 junction is cleaved by the protease *in vitro* and may represent the natural cleavage site. The RT/RNase H activities associated with p66 and p51 and p66/p51 heterodimers were determined after renaturation of these proteins from polyacrylamide gels. The p66/p51 heterodimers exhibit about eightfold higher RT and RNase H activities than each of the individual subunits alone suggesting that heterodimers are of biological relevance. Protease activity was monitored by three assays, by cleavage of denatured ovalbumin, of the MS2-*gag* fusion protein or synthetic peptides. Several synthetic peptides representing various potential protease cleavage sites and modifications thereof were analyzed *in vitro* for their efficiency of cleavage by the protease. One synthetic peptide representing the natural amino-terminal cleavage site of the protease was modified by several amino acid substitutions in order to characterize the specificity of the protease. Several of the peptides as well as cerulenin and acetyl-pepstatin were analyzed as protease inhibitors.

RESULTS

HIV-1 protease generates RT/RNase H heterodimers.

The *pol*-gene of HIV-1 codes for the p9 protease, the p66 RT/RNase H and the p32 integrase. The p66 RT/RNase H molecule if expressed by recombinant DNA technology in bacteria is either rather stable or undergoes rapid processing to a p66/p51 heterodimer, depending on the expression system and the host cell used (Di Marzo Veronese *et al.*, 1986; Farmerie *et al.*, 1987; Larder *et al.*, 1987; Le Grice *et al.*, 1987; Lightfoote *et al.*, 1986; Tanese *et al.*, 1988). If purified from virus particles, the normally obtained RT preparation consists of the heterodimeric p66/p51 form with roughly equal molar amounts of both components (Hansen *et al.*, 1987). The p51 protein is a processing product of the p66 protein from which a p15 carboxy-terminal portion is deleted (Hansen *et al.*, 1988b). While the p66 protein exhibits RT and RNase H activity in activated gel analysis (Hansen *et al.*, 1988b), the p51 shows a significantly reduced RT activity which is often undetectable (Hansen *et al.*, 1988b; Lori *et al.*, 1988). It completely lacks RNase H activity. The carboxy-terminal cleavage product p15 when isolated from HIV-1 particles exhibited strong RNase H activity (Hansen *et al.*, 1988b). In previous studies synthetic heptapeptides representing putative cleavage sites of *gag* and *pol* proteins proved useful as substrates of the partially purified p9 protease *in vitro* (Billich *et al.*, 1988). A peptide from the p51/p15 junction, LEKE'PIV, which had been selected because of some similarity to the p17/p24 *gag* junction resisted to protease cleavage *in vitro* (Billich *et al.*, 1988). Based on experience with the heterodimeric RT from avian myeloblastosis virus (AMV) (Moelling, 1974a,b), we assumed that the HIV-1 protease was responsible for the p66/p51 transition. In the case of AMV not only the AMV specific protease p15, but trypsin and other proteases, also allowed transition of the larger to the smaller subunit suggesting the existence of a site sensitive to proteases in general. A similar situation can be envisaged with HIV-1. Lowe *et al.* (1988) demonstrated limited proteolysis of p66 by α-chymotrypsin. Based on sequence comparison, Johnson *et al.* (1986) had pointed out to the two domains of the p66 molecule, the RT and RNase H, linked by a flexible 'tether' region which may be the target of more than one protease.

We investigated whether treatment of the p66 RT/RNase H by the p9 protease *in vitro* generates a p51 and the previously described p15 (Hansen *et al.*, 1988b). A ptac-*pol* expression system according to published procedures (Larder *et al.*, 1987) was used which was kindly supplied to us by Dr. G. Tarpley , USA. The RT was purified by DEAE-phosphocellulose and polyU-Sepharose as published (Hansen *et al.*, 1987). The recombinant protease p9 was expressed and partially purified as published (Billich *et al.*, 1988; Hansen *et al.*, 1988a). One to two micrograms of p66 and 5 to 10 micrograms of the protease were incubated in a standard protease assay at 37°C. The reaction products were analyzed by Western blot using a monoclonal antibody (MAB) 5/23 as described (Hansen *et al.*, 1988b). As can be seen in Figure 1A treatment of p66 with p9 protease generates a p51 and a p15 from p66. The period of incubation ranged from 10 min to 14 hrs with no significant difference suggesting resistance to the protease of the p66/p51 heterodimer. The p15 protein exhibited RNase H activity in activated gel analysis (Moelling and Schulze, manuscript submitted).

In order to identify the exact cleavage site between the RT and RNase H domains, we chose another heptapeptide AETF'YVD as potential substrate because of its similarity with other cleavage sites of *gag* and *gag-pol* with an F at the -1 position. Treatment of this peptide by the p9 protease *in vitro* resulted in cleavage which was monitored by subsequent analysis of the reaction products by HPLC as described (Billich *et al.*, 1988). Analysis of the amino acid composition of the two cleavage products indicates cleavage between F and Y (Figure 1B). This site corresponding to amino acid residue 437 of the p66 RT may represent a natural p9 protease cleavage site *in vivo*.

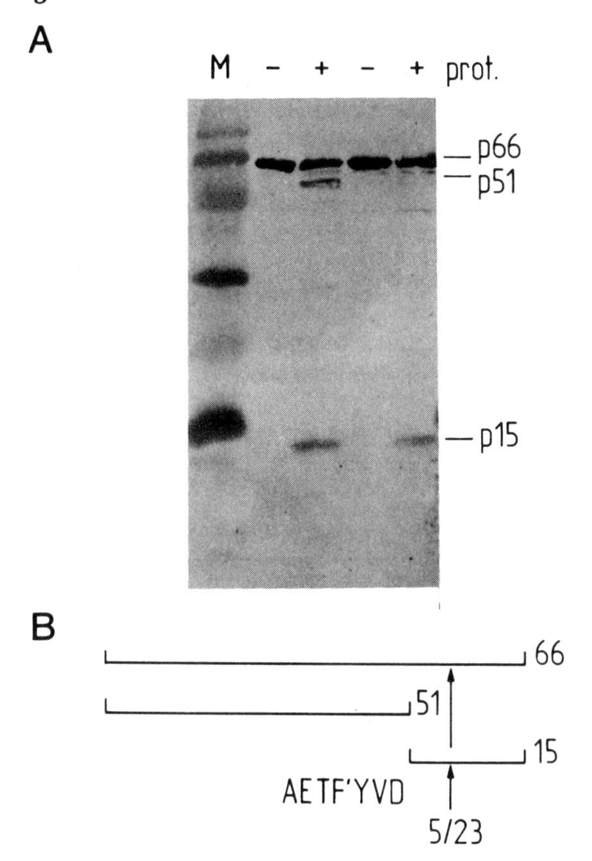

C

RT / RNASE H Activities

proteins	RT	RNase H
p66	17.857	4.690 (6 %)
p66/p66	41.237	26.630 (35 %)
p51	5.027	3.300 (4 %)
p51/p51	6.922	background
p66/p51	165.265	41.840 (65 %)
		75.840 (100 %)

Figure 1.

Cleavage of p66 by p9 protease *in vitro*. (A) In a standard protease test 1 ml of RT/RNase H purified from ptac-*pol* expressing bacteria, which consisted mainly of p66 and an intermediate protein slightly larger than p51, was incubated in the presence (+) or absence (-) of the partially purified recombinant HIV-1 p9 protease (prot.) (Billich et al., 1988) at 37° C for 1 hr. The reaction products were separated on a 10% SDS-polyacrylamide gel and processed for immunoblotting using a MAB 5/23 against the RNase H and another MAB against p51 (Hansen *et al.*, 1988b). M indicates marker proteins (92,68,45,28,18 kDa from top to bottom). (B) A schematic drawing illustrates the reaction seen in Fig. 1A. AETF'YVD is a heptapeptide cleaved by the protease *in vitro* which may represent the cleavage site between p51 and p15. (C) The p66 and p51 proteins were cut out of gels and slowly renatured either individually or together according to Hager and Burgess (1980). RT and RNase H activities were measured by standard assays (Hansen et al., 1987). The radioactive hybrid used as input was defined as 100%. The numbers indicate the amount of cpm released from the hybrid and reflect the activity of the RNase H, which is also given in %.

The natural occurence and stability of the p66/p51 heterodimer raised the question whether this form is of an advantage not only against proteolytic at tacks but also in terms of its associated enzymatic functions. To investigate this possibility the p66 and p51 proteins from a heterodimeric RT/RNase H preparation were recovered from SDS-polyacrylamide gels and submitted to a renaturation process according to Hager and Burgess (1980). The p66 and p51 preparations were renatured individually and also combined whereby only a molar ratio of about 1 to 0.2 for p66 to p51 (not 1 to 1) was accomplished because of shortage of material for p51. The result of standard RT and RNase H assays using p66, a p66/p66 mixture, p51, a p51/p51 mixture, and a p66/p5 1 heterodimer is shown in Figure 1C. The RT activity of the p66/p51 heterodimer is about 8-fold higher than the sum of the activities of the two subunits (165,265 vs 22,884 cpm). The RNase H activity was determined using a heteropolymeric ^{35}S -labeled hybrid as described (Hansen *et al.*, 1987). The amount of hybrid input was 75,840 cpm defined as 100%. RNase H activity of the p66/p51 heterodimer amounted to hydrolysis of 41,840 cpm corresponding to 65 % of the hybrid, whereas the sum of hydrolysis by p66 and p51 amounts to only 7,990 cpm, corresponding to 10% hydrolysis of the hybrid. Since hybrid hydrolysis is not linear up to 65% one cannot exactly quantitate the stimulation of the RNase H activity by the heterodimer compared to the two subunits alone, but it is significant and severalfold.

Synthetic peptides as inhibitors of the HIV-1 protease

The HIV-1 protease is part of the *pol* precursor and liberates itself by autocatalytic processing in a time dependent fashion. In the trp9-*pol* expression system (Hansen *et al.*, 1988a) the carboxyterminus is cleaved first followed by cleavage of the amino-terminus (SFNF'PQI). The protease tends to dimerise as can be judged from gel filtration analysis (Billich *et al.*, 1988). It cleaves bacterially expressed MS2-*gag* proteins at expected cleavage sites (Hansen *et al.*, 1988a). A series of synthetic peptides have been used to characterize the specificity of the cleavage sites (Billich *et al.*, 1988; Moelling *et al.*, 1989). These analyses were performed using heptapeptides which were analyzed after treatment with the partially purified p9 protease *in vitro* by HPLC analysis (Billich *et al.*, 1988). We used a series of synthetic peptides to test their abilities to inhibit a synthetic heptapeptide which has been used as a model substrate before. It corresponds to the cleavage site between p17 and p24 *gag*, SQNY'PIV (Billich *et al.*, 1988). The efficiency of cleavage of this model peptide by the protease in the presence of inhibitors can be determined by calculating the amount of substrate left after the reaction in comparison to the substrate input which corresponds to 100%. The result of such an analysis of the inhibitory effect of several synthetic peptides is shown in Figure 2A. The peptides are numbered and listed in Figure 2B. The origin of the sequences is indicated to the right of the Table . The expected cleavage sites of these peptides had been modified. In three cases (1,4,9) the bond at the cleavage site was reduced and in the residual 7 other cases the amino acid statine was introduced. Statine acts as a non-hydrolyzable dipeptide analogue resistant to attack of proteases (Kay *et al.*, 1987). While peptides 5 and 7 did not show much of an inhibitory effect on cleavage of the SQNY'PIV substrate, 3,8,9,10 and 11 were efficient inhibitors (inhibition curve of 8 not shown). The amount of inhibitor which results in 50% inhibition is derived from the curves shown in Figure 2A and listed in Figure 2B. It is around or below 50 μM for these five inhibitors.

Importance of the conformation of substrates

It needs to be shown whether it is allowed to generalize from inhibition of cleavage of a synthetic heptapeptide to other substrates. Synthetic substrates are due to their inherent flexibilities most likely pressed into a suitable conformation to fit into the cleft of the protease. This may not be possible with larger substrates. We noted that cleavage of the *gag* protein, which harbors several natural cleavage sites of the HIV-1 protease, depended not only on sequence but also on conformation. The *gag* protein can be completely resistant if denatured *in vitro* e.g. after SDS-polyacrylamide gel electrophoresis (Hansen *et al.*, 1988a). On the contrary, the p66/p51 heterodimer which was mentioned above to be resistant to protease treatment, can be completely digested after mild treatment with detergent, e.g. 0.01% SDS (Schulze and Moelling, unpublished observation). We had described for the AMV p15 protease, that various substrates are accessible to its proteolytic attack only after partial denaturation (Dittmar and Moelling, 1978). Here we used a similar approach for the HIV-1 protease. The protein ovalbumin completely resists treatment to the HIV-1 protease *in vitro* (Figure 2C, C4) in the absence of SDS or in the presence of SDS below 0.01 and above 0.1% (not shown). Only at an SDS concentration around 0.004% is the ovalbumin sensitive to protease treatment (Figure 2C, C1 and C3). This concentration of SDS was used in standard ovalbumin tests. Inhibitors 3,5,7,8 and 9 were added to the protease assay at the molar concentrations indicated (0.1 to 0.4). While 8 and 9 strongly inhibit proteolysis of ovalbumin, 7 is least inhibitory. Cleavage can be detected by reduction of the intensity of the ovalbumin band itself as well as the appearance of distinct cleavage products. Inhibitors 3 and 5 only exhibit slight inhibitor effects, inhibitor 3 prevents appearance of the smallest degradation product migrating close to the front of the gel. This assay, based on cleavage of a partially denatured protein, can serve as screening test for the detection of inhibitors of the HIV-1 protease . Its result is consistent with most of the results shown in Figure 2A, except compound 3 behaves differently in the two tests. It is impossible to judge which substrate most closely mimmicks the natural situation.

One can envisage cleavage of a protein involved in some chromophoric reaction in order to determine the protease effect by scoring the absorbance similar to ELISA assays. It is noteworthy that the HIV-1 protease does not completely digest this substrate. It apparently recognizes specific sequences of the ovalbumin and cleaves only at a few specific sites - suggesting a strong selection for cleavable sites. These sites of the ovalbumin are presently analyzed by amino-acid sequencing. The HIV-1 protease bears similarity to the V8 protease from *S. aureus* which proved to be useful in generating partial proteolytic fragments for comparison of proteins (Cleveland *et al.*, 1977).

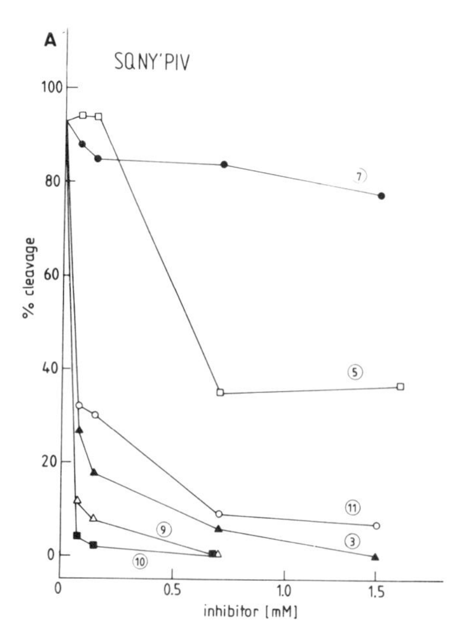

B

#	Inhibitor	50 % inhibition	Comment
1	VSFN-phe^{red}pro—QITL	< 125 μM	N-term HIV-1 prot
2	VSFN—phe—sta—QITL	< 125 μM	N-term HIV-1 prot
3	VSFN—cys—sta—QITL	~ 50 μM	N-term HIV-1 prot
4	CTLN—phe^{red}pro—ISPI	125 μM	HIV-1 prot/RT
5	CTLN—phe—sta—ISIP	550 μM	HIV-1 prot/RT
6	TATI—sta—QRGN	< 125 μM	HIV-1 gag p24/p15
7	ERQA—cys—sta—LGKI	>>1.5 mM	HIV-1 gag/pol
8	RGLA—sta—QFSL	<< 50 μM	HIV-2
9	VSFN—phe^{red}pro—QITL	~ 50 μM	HIV-N-term prot
10	VSLN—phe—sta—QVSQ	~ 50 μM	HIV-1 ideal substrate
11	MSLN—sta—VAKV	~ 50 μM	HIV-2

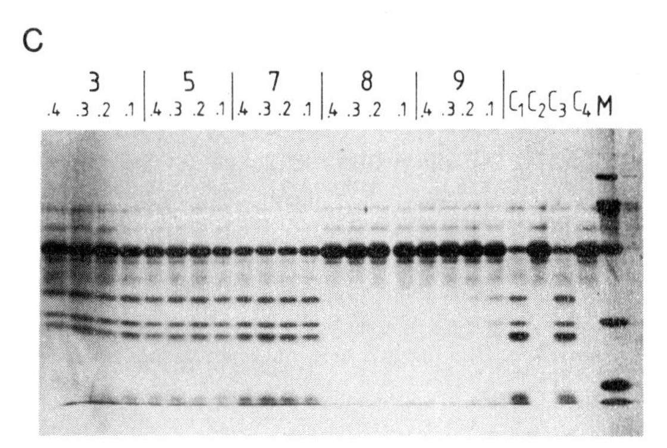

Figure 2.

(A) The heptapeptide SQNY'PIV, representing the p24/p17 junction is efficiently cleaved by the partially purified protease *in vitro*. The protease assay was inhibited by a series of synthetic peptides listed in (B) and numbered 1 to 11. The reaction products were analyzed by HPLC and the amount of cleavage calculated by determination of the areas of the substrate S and the two cleavage products P1 and P2 (Billich *et al.*, 1988). The circled numbers represent the numbers of the list of compounds shown in (B). (C) shows the result of cleavage of ovalbumin as substrate S (5 micrograms) which was performed with the partially purified protease under standard assay conditions except that 0.004% SDS was present during the reaction. The reaction products are analyzed by gel electrophoresis (10%) and Coomassie blue staining. The inhibitors indicated by numbers (see (B) for sequences), were present throughout the reaction. The concentrations ranged from 0.1 to 0.4 mM. C1 and C3 represent controls in the presence of protease and absence of inhibitors, C2 and C4 show stability of S with and without incubation. M represents marker proteins (see Fig. 1A).

Modifications of one cleavage site

In order to learn more about the sequence specificity of the HIV-1 protease we chose one of the model substrates, SFNF'PQI, representing the amino-terminus of the protease itself. This substrate is cleaved efficiently. The amino-acids at positions +1 and -1 were modified as shown in Figure 3A. Replacement of the unpolar and uncharged P by G (polar and uncharged), K (polar and positively charged), or D (polar and negatively charged) abolished the property of the peptide to serve as substrate as did replacement at the -1 position F (unpolar and uncharged) by L (unpolar, uncharged), H (polar and positively charged), K (polar and positively charged) and D (polar and negatively charged). It must be concluded that the F'P junction is vitally important for this particular peptide substrate.

It had been proposed by Pearl and Taylor (1987) that the substrate of the HIV-1 protease would recognize a sequence XY'PZ with X being small and hydrophobic, Y aromatic or large and hydrophobic, and Z small and hydrophobic. The predicted cleavage site between an aromatic amino acid and the proline was confirmed by the peptide described here, SFNF' PQI and the abovementioned SQNY'PIV as well as a third one described

A

#	Sequence of substrate	Cleavage
1	SFNF' PQI	yes
2	SFNF' GQI	no
3	SFNF' KQI	no
4	SFNF' DQI	no
5	SFNL' PQI	no
6	SFNH' PQI	no
7	SFNK' PQI	no
8	SFND' PQI	no

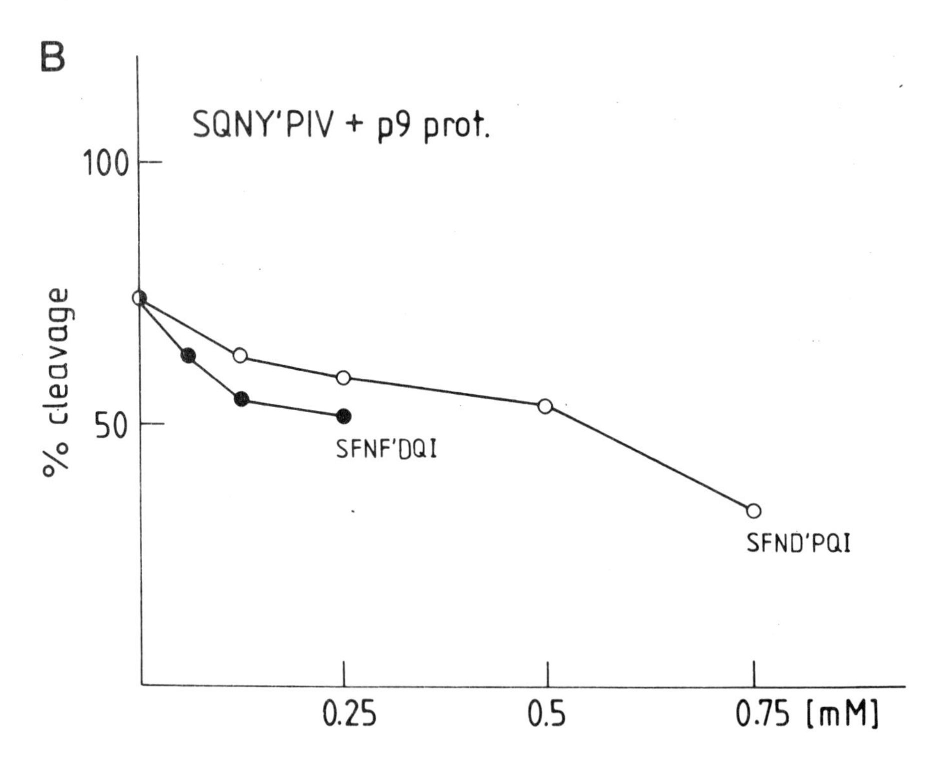

Figure 3.
(A) The heptapeptide SFNF'PQI which represents the amino- terminus of the protease cleavage site, is an efficient substrate of the partially purified protease *in vitro*. At the positions +1 and −1 from the cleavage site (indicated by hiatus) several amino acid exchanges have been introduced (numbers 2 to 8), all of which result in resistance to cleavage. (B) Compounds 4 and 8 were tested as examples for their effect on cleavage of the heptapeptide SQNY'PIV and the p9 protease. The assay conditions have been described (Billich *et al.*, 1988).

earlier, TLNF'PIS (Billich *et al.*, 1988). However, we have analyzed other cleavage sites, which do not contain the proposed consensus sequence and are cleaved as well, e.g. ARVL'AEV (p24/p15), ATIM'MQR (p24, p15), and RKL'FLD (p66/p32) (Billich *et al.*, 1988).

Those peptides which resisted cleavage were tested for their ability to inhibit cleavage of another substrate, e.g. SQNY'PIV and cleavage was again determined by analysis of the substrate by HPLC chromatography. SFNF'DQI and SFND'PQI were tested in such a way and exhibited low inhibitory effects (50% cleavage above 250 μM) (Figure 3B).

Inhibitory compounds: acetyl-pepstatin and cerulenin

We have analyzed the effect of pepstatin A which is a known inhibitor of pepsin and other aspartic proteases and also of HIV-1 protease (e.g. Hansen et al., 1988a) in comparison with its acetylated form, acetyl-pepstatin. We used an MS2-*gag* fusion protein as substrate for the HIV-1 protease *in vitro* and tested the result of cleavage by immunoblotting with a MAB against the MS2 portion which is fused after cleavage to p17, MS2-p17. Details of the assay have been described (Hansen *et al.*, 1988a). The result is shown in Figure 4A which indicates that both, pepstatin A as well as acetyl-pepstatin A inhibit the protease at 0.1 mM concentrations with no significant differences among the two compounds.

Another compound cerulenin (2,3-epoxy-4-oxo-7,10-dodecadienoylanide) has been described as inhibitor which significantly affects the production of infe ctious Rous sarcoma virus (Goldfine *et al.*, 1978) and murine leukemia virus (MuLV) by cells infected in culture (Ikuta and Luftig, 1986, Ikuta *et al.*, 1986), with an apparent decrease in specific proteolytic cleavage of the Pr65*gag* precursor protein possibly due to inhibition of the viral protease. Recently, Pal *et al.* (1988) reported a similar effect on precursor processing of HIV-1. Originally cerulenin has been described as an inhibitor of fatty acid and sterol synthesis (D'Angelo *et al.*, 1973) which inhibits the fatty acylation of the glycoproteins of vesicular stomatitis virus (Schlesinger and Malfer, 1982). We used cerulenin to analyze its inhibitory effect on the HIV-1 protease with the MS2-*gag* substrate. As can be seen in Figure 4B, cerulenin inhibits most of the cleavage at a concentration at 0.1 mM and higher (Moelling *et al.*, 1990). Cerulenin most likely binds to the active site of aspartic proteases. The presence of an epoxide function in cerulenin is reminiscent of a similar function in a known aspartic protease inhibitor EPNP (Tang, 1971), which irreversibly inactivates aspartic proteases by esterification of the two active site aspartic acid residues (James *et al.*, 1977). Cerulenin itself is cytotoxic and therefore inappropriate for clinical use. It provides, however, a lead for the rational design of active-site-directed protease inactivators.

ACKNOWLEDGEMENT

This work was supported by a grant from the Bundesministerium für Forschung und Technologie, number FKZ II-019-86.

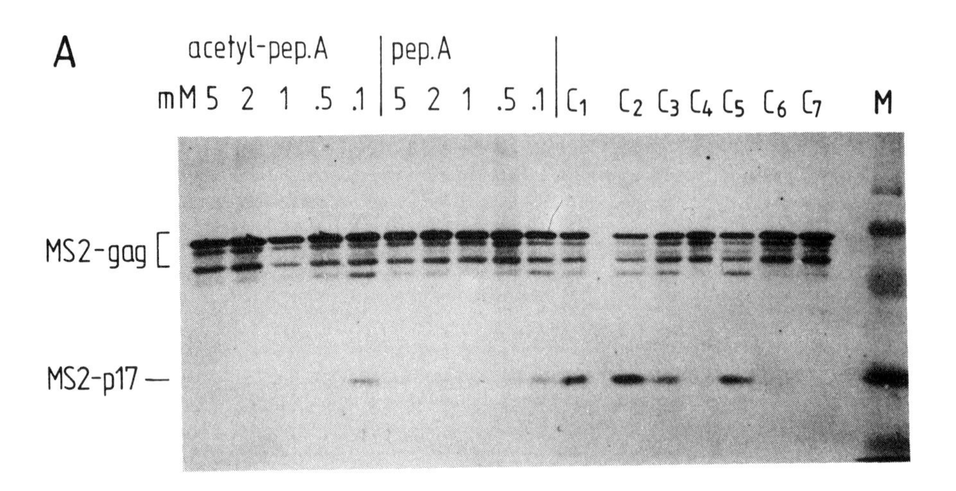

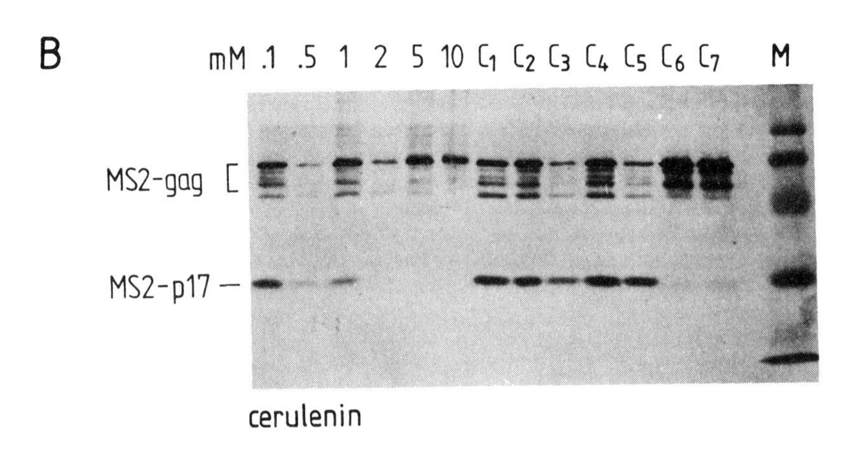

Figure 4.
(A) The MS2-*gag* fusion protein expressed in bacteria was present in a bacterial lysate and was incubated as substrate with the partially purified protease as described (Hansen *et al.*, 1988a). The cleavage product was monitored through its MS2 moiety by using a MAB against MS2 in a Western blot (Hansen *et al.*, 1988a). Acetyl-pepstatin and pepstatin A at 0.1 to 5 mM concentrations were added to the reaction as inhibitors. C1 to C7 are controls excluding effects by buffers or incubation time. M represents a molecular weight marker (68,45,28,18 kDa from top to bottom). (B) Same experiment as in (A) except that cerulenin was used as an inhibitor (0.1 to 10 mM). The protease and cerulenin were incubated for 30 min at room temperature before the protease assay. Other details as in (A) (Moelling et al., 1990).

REFERENCES

Billich,S., Knoop,M.-T., Hansen,J., Strop,P., Sedlacek,J., Mertz,R., and Moelling,K. *J.Biol.Chem.*, **263**, 17905-17908, (1988).

Cleveland,D.W., Fischer,S.G., Kirchner,M.W., and Laemmli,U.K. *J.Biol.Chem.*, **252**, 1102-1106, (1977).

D'Angelo,G., Rosenfeld,I.S., Awaya,J., Omura,S. and Vagelos,P.R. *Biochim.Biophys.Acta*, **326**, 155-166, (1973).

Di Marzo Veronese,F., Copeland,T.D., DeVico,A.L., Rahman,R., Orozlan,S., Gallo,R.C. *Science*, **231**, 1281-1291, (1986).

Dittmar,K.J. and Moelling,K. *J. Virol.*, **28**, 106-118, (1978).

Farmerie,W.G., Loeb,D.H., Casavant,N.C., Hutchinson III,C.A, Edgell,M.H. and Swanstrom,R. *Science*, **263**, 305-308, (1987).

Goldfine,H., Harley,J.B. and Wyke,J.A. *Biochim.Biophys.Acta*, **152**, 229-240, (1978).

Hager,D.A. and Burgess,R.R. *Anal. Biochem.*, **109**, 76-86, (1980).

Hansen,J., Schulze,T., and Moelling,K. *J.Biol.Chem.*, **262**, 12393-12396, (1987).

Hansen,J., Billich,S., Schulze,T., Sukrow,S., and Moelling,K. *EMBO J.*, **7**, 1785-1791, (1988a).

Hansen,J., Schulze,T., Mellert,W. and Moelling,K. *EMBO J.*, **7**, 239-243, (1988b).

Ikuta,K. and Luftig,R.B. *Virology*, **154**, 195-206, (1986).

Ikuta,K., Coward,J. and Luftig,R.B. *Virology*, **154**, 207-213, (1986).

James,M.N.G., Hsu,I.-N. and Delabaere,L.T.J. *Nature*, **267**, 808-813, (1977).

Johnson,M.S., McClure,M.A., Feng,D.F., Gray,J. and Doolittle,R.F. *Proc.Natl.Acad.Sci. USA*, **83**, 7648-7652, (1986).

Kay,J., Jupp,R.A., Norey,C.G., Richards,A.D., Reid,W.A., Taggart,R.T., Samloff,J.M. and Dunn,B.M. in 'Proteases: Potential Role in Health and Disease' (Heidland, A. and Horl, W.H., eds.). Plenum Press, (1987).

Larder,B., Purifoy,D., Powell,K. and Darby,G. *EMBO J.*, **6**, 3133-3137, (1987).

LeGrice,S.F.J., Beuck,V. and Mous,J. *Gene*, **55**, 95-103, (1987).

Lightfoote,M.M., Coligan,J.E., Folks,T.M., Fanci,A.S., Martin,M.A. and Venkatesan,S. *J.Virol.*, **60**, 771-775, (1986).

Lori,F., Scovassi,A.I., Zella,R., Achilli,G., Cattaneo,E., Casoli,C. and Bertazzoni,U. *AIDS Res. Hum. Retro.*, **4**, 393-398, (1988).

Lowe,D.M., Aitken,A., Bradley,C., Darby,G.K., Larder,B.A., Powell,K.L. Purifoy,D.J.M., Tisdale,M. and Stammers,D.K. *Biochemistry*, **27**, 8884-8889, (1988).

Moelling,K. *Cold Spring Harbor Symp.Quant. Biol.*, **39**, 969-973, (1974a).

Moelling,K. *Virology*, **62**, 46-59, (1974b).

Moelling,K., Knoop,M.-T., Billich,S., Blaha,J., Pavlickova,L. and Soucek,M. in 'Proteases and Retroviruses'. Proceedings of the 14th Int. Congress of Biochemistry, (V. Kostka, ed.), W. de Gruyter and Co., Berlin, p.155-164, (1989).

Moelling,K., Shulze,T., Knoop,M.-T., Kay,J., Jupp,R., Nicolaou,G. and Pearl,L.H. *FEBS Lett.*, in press, (1990).

Pal,R., Gallo,R.C. and Sarngadharen,M.G. *Proc.Natl.Acad.Sci.USA*, **85**, 9283-9286, (1988).

Pearl,L.H. and Taylor,W.R. *Nature*, **328**, 482, (1987).

Prasad,V.R. and Goff,S.P. *Proc.Natl.Acad.Sci.USA*, **86**, 3104-3108, (1989).

Schlesinger, and Malfer, *J.Biol.Chem.*, **257**, 9887-9890, (1982).

Tanese,N., Prasad,V.R. and Goff,S.P. *DNA*, **7**, 407-416, (1988).

Tang,J. *J.Biol.Chem.*, **246**, 4510-4517, (1971).

4
Retroviral Protease: Substrate Specificity and Inhibitors

Yoshiyuki Yoshinaka, Iyoko Katoh and Kohei Oda

RETROVIRUS PARTICLE ASSEMBLY

Protease activity is involved in particle formation and/or the maturation process of various virus systems: structural and/or envelope proteins are synthesized as polyprotein precursors and assembled to form immature particles. Those precursors are processed into functional components by proteinases. This type of assembly may be more accurate and economical than a system where particles are formed by assembling separate proteins with distinct properties, affinities and functions. Viruses of picorna-, toga-, flavi-, adeno-, and retrovirus families have specific proteinases, that is, their viral genomes encode proteinases. The enzymes work as a processing enzyme which cleaves viral polyprotein precursors into mature components. The precursor processing is essential for production of biologically active virus particles (Klausslich and Wimmer, 1988).

A typical example of the protease-dependent particle maturation was demonstrated in Moloney murine leukaemia virus (MuLV). When the immature cores were isolated and treated with the specific protease, their structure was converted to mature form. The morphological conversion was accompanied by specific cleavage of the precursor polyprotein, a major component of the immature cores (Yoshinaka and Luftig, 1977).

Based on the observations and other pieces of evidence accumulated in molecular biological studies, an assembly model of retrovirus particles was presented as follows: All replication competent retroviruses contain three essential genes, *gag*, *pol* and *env*. The *gag* and *pol* genes encode core structural proteins and enzymes, respectively (Varmus, 1988). The precursor-processing enzyme called retrovirus protease (PR) is encoded by the 5' terminus of the *pol* gene (von der Helm, 1977; Yoshinaka *et al.*, 1985a, 1985b). Translation of genome-sized mRNA generates the *gag* gene product and the readthrough product of *gag-pol* genes. The *pol* proteins starting from PR are synthesized employing either a readthrough or frameshift suppression mechanism of the *gag* termination codon (Yoshinaka *et al.*, 1985a, 1985b; Jacks *et al.*, 1988). The *gag* and *gag-pol* precursor polyproteins thus synthesized are directed to the cellular membrane because of the NH_2-terminal fatty acid modification (Henderson *et al.*, 1983), where they assemble and package the genome RNA to form particles. Probably during the particle budding precursor processing begins with autoprocessing of the *gag-pol* precursor molecules. Functional components of the core structures and fully-active enzymes are generated. Particles are rebuilt in an active form. This maturation event is essential for virus infectivity (Katoh *et al.*, 1985; Kohl *et al.*, 1988).

In addition to this particle formation process, the molecular association of retrovirus components of the mature particles has long and throughly studied with interesting results. As shown in Table 1, most of the viral components have been demonstrated to self-associate and form dimers: the virion has a diploid genome. Murine retrovirus *gag* precursor polyprotein Pr65 was detected as a dimer in the absence of reducing reagent (Yoshinaka *et al.*, 1984). Dimers of the mature *gag* proteins, including MA, phosphoprotein, CA, and NC, have been demonstrated by cross-linking experiments in various retrovirus systems (Pinter and Fleissner, 1979; Dittmar *et al.*, 1980; Uckert *et al.*, 1982). PR exhibits the activity by forming a dimer (Katoh *et al.*, 1989; Meek *et al.*, 1989). The surface glycoprotein of Friend spleen focus forming virus has recently been demonstrated to occur as dimers (Gliniak and Kabat, 1989).

Table 1.
Molecular association of retrovirus components.

Component	Dimer formation
genome RNA	yes
gag proteins	
gag precursor polyprotein	yes
MA	yes
phosphoprotein	yes
CA	yes
NC	yes
pol proteins	
PR	yes
RT	(yes)*
IN	?
env proteins	
SU-TM fusin protein**	yes
SU	?
TM	?***

?, dimer formation not demonstrated; *, data not available in publications; ** SU-TM fusion protein of Friend spleen focus forming virus; ***, highly aggregative. References are cited in the text.

These observations suggest that the basic unit of molecular organization in retrovirus particles is homodimers of each component; It is possible that they are functional in a dimeric form. Highly organized structure of the virus particles would be constructed by further interactions between homologous homodimers and also among heterologous homodimers.

It thus appears to be advantageous for the virus to encode halves of the several functional units and then make them fully active by dimerization following protein synthesis. In this sense, the virus genome is well organized and simplified. Furthermore, using this strategy, virus provides stability of the genetic information.

CHARACTERIZATION OF PROTEASES

We purified PRs from virus preparations of several origins including murine, feline, bovine and avian. Viral proteins were first precipitated with acetone and then extracted with NaCl (1M) at pH 7.8. The extract was fractionated by Sephadex G-75 column chromatography. Fractions containing the protease activity were pooled and completely solubilized with 6M guanidine hydrochloride at pH 3.0 for subsequent analysis by reverse phase-high performance liquid chromatography. The protease was purified to make a single band on sodium dodecylsulfate-polyacrylamide gel electro phoresis (SDS-PAGE) gels (Yoshinaka *et al.*, 1985a, 1985b).

To characterize retrovirus PRs biochemically, we used Moloney murine sarcoma virus (MSV)-derived particles (Gazdar strain) (Gazdar et al, 1971) as the substrate. Gazdar MSV is defined as a naturally occurring protease-deficient virus, and therefore contains *gag* precursor Pr65 as the core structural protein (Yoshinaka and Luftig, 1982). Our assay system included leupeptin which suppresses serum-derived protease activity contaminated in the Gazdar MSV preparations. The reactions were at low ionic strength and at pHs varied depending on the origin of the virus protease. After incubation for 16hrs at 22°C, the protein composition was analyzed by SDS-PAGE. Protease activity was determined by a decrease of Pr65-band intensity and concomitant increase of specific cleavage products. Roughly one molecule of AMV protease can made one cut in 20 molecules of Pr65 at pH 6.0.

In this assay system, the protease activity is inhibited (up to 50%) by high salt concentrations or ionic strength (>0.1M NaCl or KCl), and divalent cations(>0.05M Mg^{2+} or Mn^{2+}). Neutral sugar, non-ionic detergents and DMSO do not affect the activity.

pH optima were examined in BLV, avian myeloblastosis virus (AMV) and human immunodeficiency virus (HIV) PRs (Figure 2). AMV PR was most active at pH 6.0 to 6.5. BLV protease exhibited its activity in a relatively wide pH range (from pH 4.5 to 8.0). HIV PR has the pH optimum in a neutral range between 6.5 to 7.0. Murine and feline leukaemia virus PRs also have their pH optima at pH 7.0 to 7.5.

Interestingly, as observed most clearly in the BLV PR, the cleavage specificity was shifted as the pH condition changed from acidic to neutral range. HIV PR also produced a new product at lower pHs. These

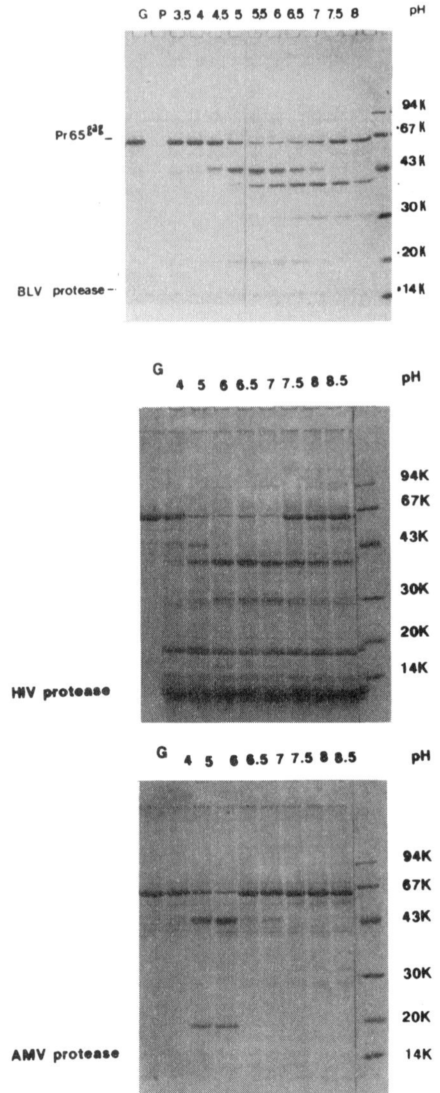

Figure 1.
Retrovirus PR activity under various pH conditions. BLV (top), HIV (middle) and AMV (bottom) PRs were examined for cleavage of mouse *gag* precursor Pr65. Assay conditions are described in the text. G and P indicate substrate Gazdar MSV and PR, respectively.

results suggest that the specificity of PRs is affected by the conformational changes of the substrate polyprotein precursor. The cleavage pattern at the neutral pHs may represent their specificities *in vivo*.

Detailed analysis of their cleavage sites in Pr65 is described in the following sec tion.

PR has been characterized as an aspartic proteinase based on the observations listed below. (i) PRs of various origins from yeast to human have conserved amino acid sequence containing a triplet of Asp-Thr(Ser)-Gly, which is the catalytic center of aspartic proteinases (Toh *et al.*, 1985; Pearl and Taylor, 1987). (ii) PR activity is inhibited by pepstatin A, a substrate analogue inhibitor of aspartic proteinases (Katoh *et al.*, 1987). (iii) X-ray crystallographic analyses of avian and human PRs have revealed close resemblance between PR and aspartic proteinases in peptide holding (Miller *et al.*, 1989; Navia *et al.*, 1989; Wlodawer *et al.*, 1989).

Retroviral PR, however, has several properties distinctive from those of common aspartic proteinases such as pepsin, cathepsin D, renin and fungus proteinases. (i) As shown above, their optimum pHs were found in a range from 6.0 to 7.5 by our assay system using natural substrate Pr65. (ii) Limited native proteins are available for their substrate. Only *gag-* and *pol-* related polyprotein precursors can serve as the viral enzyme's substrate without denaturation. (iii) PR molecules self-associate to form a dimer, which corresponds to one molecule of common aspartic proteinases. Aspartic proteinases are two-domain enzymes: both NH_2- and COOH- terminal domains have the active site, Asp-Thr-Gly, and form a catalytic center in a roughly symmetrical conformation. PR molecules have a single Asp-Thr-Gly sequence and are 99 to 125 amino acids long, half the size of aspartic proteinases. Several lines of evidences have been obtained in our experiments, including molecular weight estimation by gel filtration and analytical centrifugation, cross-linking with bifunctional reagent and competitive inhibition of the activity by the peptides obtained by CNBr-cleavage of the protease. These suggest that dimer formation is required for their activity (Katoh *et al.*, 1989). The dimerization was also evident in crystal analyses of Rous sarcoma virus and HIV PRs.

Taken together, retrovirus PRs form a subgroup of aspartic proteinases.

COMPARISON OF SUBSTRATE SPECIFICITY OF PROTEASES FROM VARIOUS ORIGINS

The substrate specificity of PRs has been extensively studied in many laboratories, but as yet not adequately explained.

We compared the substrate specificity of PRs from several species using murine *gag* precursor Pr65 as a common substrate. Based on the NH_2- and COOH- terminal amino acid sequences of MuLV mature *gag* proteins, p15, p12, p30 and p10, and analyses of the intermediate cleavage products, correct processing of Pr65 by MuLV PR was described in detail (Oroszlan and Gilden, 1980). Although MuLV PR cleaves the *gag* precursor at three positions (Figure 2, positions 1, 4 and 6), cleavage at position 4 precedes cleavage at position 1 or 6. The processing occurs in order but not randomly.

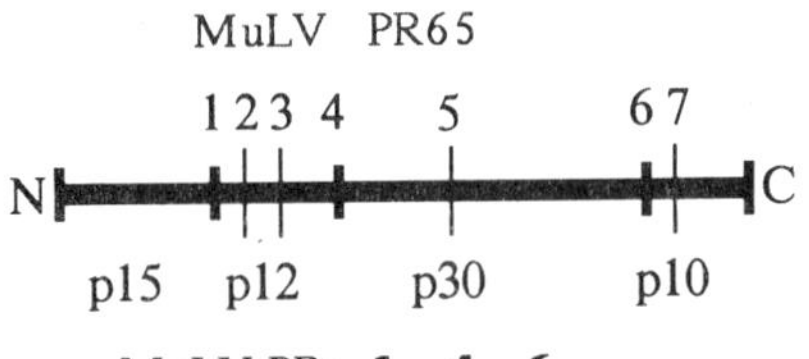

MuLV PR: **1. 4. 6.**

AMV PR: **2. 4. 5**(ANS).
BLV PR: **3**(acidic). **4.** 6.
HIV PR: **1. 4.** 6. 7.

	P 5	P 4	P 3	P 2	P 1		P'1	P'2	P'3	P'4	P'5
1.	R	S	S	L	Y	+	P	A	L	T	P
4.	T	S	Q	A	F	+	P	L	R	A	G
6.	M	S	K	L	L	+	A	T	V	V	S
2.	G	P	L	I	D	+	L	L	T	E	P
3.	L	I	D	L	L	+	T	E	D	P	P
5.	P	N	E	V	D	+	A	A	F	P	L

Figure 2.

Cleavage sites in Pr65 by PRs from various origins. Pr65 was digested with AMV, BLV or HIV PR for determination of the cleavage sites by analysis of the products. Specificity of MuLV PR is described by Oroszlan and Gilden (1980). Positions indicated by bold-face numbers were determined by NH_2-terminal amino acid sequence analysis. Postion indicated by outlined number was determined by SDS-PAGE followed by immunoblot analysis. Other cleavage sites were based on size analysis of the product.

First, we examined AMV protease p15 for the cleavage of Pr65. Lower concentrations of p15 cleaved Pr65 at position 2 (Figure 2) to produce a 47kD and a 18kD protein. With higher activity, the 47kD protein was further cleaved at position 4 to generate a 40kD protein. During the screening of inhibitor, we found that a chemical compound, 8-anilino-1-naphthalenesulfonic acid (ANS) modified the activity, that is, a new cleavage site appeared. In position 5 to make a 28kD protein; but in the absence of ANS, no further cleavage of the 47kD product was observed even at higher concentrations of the PR. This suggests that a conformational change of the 40kD protein might be caused by ANS. The NH_2-terminal amino acid sequences of the 18kD, 47kD, 40kD and 47kD proteins were determined. The results indicate that AMV protease cleaved Pr65 at a few positions different from those cleaved by MuLV protease, except for position 4. It should be noted that AMV protease was active at lower pHs, but had little activity at pH 7 even at higher concentrations of the enzyme.

BLV PR also cleaved Pr65. The cleavage pattern was very close to that of MuLV protease: 40kD, 27kD, 30kD and 10kD products were obtained. NH_2-terminal amino acid analysis of the 40kD product showed that the

BLV PR cleaved at position 4, which is at the junction between p12 and p30. Under acidic conditions, position 3 was cleavable, and a 46kD product was generated. This position found in p12 domain is very close to position 2 cleavable by AMV protease. Positions 3 and 2 are separated by only two amino acid residues. The conformation of the Pr65 molecule might be altered at low pH, which may have made the middle part of p12 domain become susceptible to these PRs. As described above, BLV PR is active under a wider range of pH conditions than other retrovirus PRs. These indicate that BLV PR has properties of both avian and mammalian retrovirus PRs.

HIV PR was expressed in *Escherichia coli* using φ-10 promoter/T7 RNA polymerase system. We partially purified the 9kD HIV PR from bacteria by acetone treatment followed by extraction and fractionation by Sephadex G-75 column chromatography. The HIV PR processed Pr65 correctly as MuLV PR. This was determined by NH_2-terminal sequence analysis of 40kD (position 4), 30kD (position 4), and 12kD (position 1) products. The product of 10kD was identified as p10 by immunoblotting after SDS-PAGE using mono- specific antisera raised against each of the MuLV *gag* proteins. In addition to these sites, another cleavage site (position 6) was present within the p10 domain. After the 40kD product was generated, HIV PR cleaved it either at the junction of p30 and p10, or at an unidentified position in p10 domain.

In summary, Pr65 can serve as the substrate of retrovirus PR from any origin so far examined. All retroviral PRs cleave Pr65 at the same position between p12 and p30 (position 4). Mammalian retroviral proteases process Pr65 correctly or nearly correctly. However, under some special conditions, such as at acidic pHs, they are able to cleave different positions.

No obvious sequence similarity was found in the stretches surrounding the cleavage sites. The conformation of polyprotein molecule may be important for the recognition of scissile peptide bond by PRs. The cleavage sites are often in short hydrophobic peptides flanked by polar residues, and are supposed to be prominent on the surface of the precursor molecule.

INHIBITION OF PROTEASE ACTIVITY *IN VITRO*

One of our purposes in this study was to find inhibitors applicable in clinical use as anti-retroviral drugs. As we described earlier (Katoh *et al.*, 1987), pepstatin A inhibits retrovirus protease activity at higher concentrations (>0.1mM) *in vitro*. As pepstatin A is hardly soluble in water, we could not test its effiency in culture systems. On the contrary, water soluble acetyl pepstatin did not inhibit PRs under our assay conditions. Synthetic peptides are now being used as substrates in many laboratories. The viral PRs exhibit a different reactivity toward short peptides: in contrast to the reaction with natural substrates, the enzymes require high salt concentrations (2M) and low pH conditions (at about pH 3) to cleave synthetic peptides. In the screening of inhibitors, it is important to consider *in vivo* circumstances where the protease works.

We wondered about the possibility of utilizing microbial products as the source of PR inhibitors. We screened fungus culture fluids for inhibition of aspartic proteinases, porcine pepsin and *Xanthomona sp.* carboxypeptidase, and isolated several hundred strains from among five thousand. Those culture fluids were then tested for their inhibitory effect against AMV protease. Eight out of 369 strains were found to produce inhibitory material and identification of the compounds is now being made. So far, a pepstatin analogue has been isolated from one of the strains.

Well-known microbial products, neomycin and neomycin-derivatives like G418 were found to inhibit AMV protease strongly *in vitro*; IC_{50} of neomycin-sulfate was $3x10^{-6}$ M. Unfortunately, they showed little inhibitory effect *in vivo* at 1mM. Neither production of immature particles nor accumulation of *gag* precursor protein was observed in MuLV, BLV, or HIV infected cell cultures. Preliminary results showed that neomycin sulfate might interact with the precursor molecules and protect them from proteolysis *in vitro*: an affinity column conjugated with neomycin bound nucleic acid-binding proteins, NCs, of AMV, BLV, and MuLV.

In conclusion, as retroviral PR is a unique enzyme and the cleavage occurs specifically in the viral precursor molecules, the processing event is a good target for the control of retrovirus proliferation. Normal processing could be prevented by blocking dimer formation, by destroying the active center, by modifying the cleavage sites, or by altering the conformation of the substrate.

The virus inhibitor could be other reagents which block particle structure formation by interfering with the association of core structural proteins, or by inhibition of the packaging of virus genome RNA into virions. Further understanding of the retrovirus assembly, infection and replication mechanisms will help us to develop a more appropriate strategy for prevention of virus proliferation.

REFERENCES

Dittmar, K.E., Brauer, D. and Moelling, K. in 'Biosynthesis, Modification, and Processing of Cellular and Viral Polyproteins', (Koch, G. and Richter, D. ed.) pp.289-299, Academic Press, (1980).

Gazdar, A.F., Phillips, L.A., Sarma, P.S., Peebles, P.T. and Chopra, H.C. *Nature*, 324, **69-72**, (1971)

Gliniak, B.C. and Kabat, D. *J. Virol.*, **63**, 3561-3568, (1989)

Henderson, L.E., Krutzsch, H.C. and Oroszlan, S. *Proc. Natl. Acad. Sci. USA*, **80**, 339-343, (1983)

Jacks, T., Power, M.D., Masiarz, F.R., Luciw, P.A., Barr, P.J. and Varmus, H.E. *Nature*, **331**, 280-283, (1988)

Katoh, I., Yoshinaka, Y., Rein, A., Shibuya, M., Odaka, T. and Oroszlan, S. *Virology*, **145**, 280-292, (1985)

Katoh, I., Yasunaga, T., Ikawa, Y. and Yoshinaka, Y. *Nature*, **329**, 654-656, (1987)

Katoh, I., Ikawa, Y. and Yoshinaka, Y. *J. Virol.*, **63**, 2226-2232, (1989)

Klausslich, H.-G. and Wimmer, E. *Ann. Rev. Biochem.*, **57**, 701-754, (1988)

Kohl, N.E., Emini, E.A., Schliet, W.A., Davis, L.J., Heimbach, J.C., Dixon, R.A.F., Scolnick, E.M. and Sigal, I.S. *Proc. Natl. Acad. Sci. USA*, **85**, 4686-4690, (1988)

Meek, T.D., Dayton, B.D., Metcalf, B.W., Dreyer, G.B., Strickler, J.E., Gorniak, J.G., Rosenberg, M., Moore, M.L., Magaard, V.W. and Debouck, C. *Proc. Natl. Acad. Sci. USA*, **86**, 1841–1845, (1989)

Miller, M., Jaskolski, M., Rao, J.K.M., Lies, J. and Wlodawer, A. *Nature*, **337**, 576–579, (1989)

Navia, M.A., Fitzgerald, P.M.D., McKeever, B.M., Leu, C.-T., Heimbach, J.C., Herber, W.K., Sigal, I.S., Darke, P.L. and Springer, J.P. *Nature*, **337**, 615–620, (1989)

Oroszlan, S. and Gilden, R.V. in 'Molecular Biology of RNA Tumor Viruses, (Stephenson, J.R. ed.) pp.299, Academic Press, (1980)

Pearl, L.H. and Taylor, W.R. *Nature*, **329**, 351–354, (1987)

Pinter, A. and Fleissner, E. *J. Virol.*, **30**, 157–165, (1979)

Toh, H., Ono, K., Saigo, K. and Miyata, T. *Nature*, **315**, 691, (1985)

Uckert, W., Westermann, P. and Wunderlich, V. *Virology*, **121**, 240–250, (1982)

Varmus, H. *Science*, **240**, 1427–1435, (1988)

von der Helm, K. *Proc. Natl. Acad. Sci. USA*, **74**, 911–915, (1977)

Wlodawer, A., Miller, M., Jaskolski, M., Sathyanarayana, B.K., Baldwin, E., Weber, I.T., Selk, L.M., Clawson, L., Schneider, J. and Kent, S.B.H. *Science*, **245**, 616–621, (1989)

Yoshinaka, Y. and Luftig, R.B. *Proc. Natl. Acad. Sci. USA*, **74**, 3446–3540, (1977)

Yoshinaka, Y. and Luftig, R.B. *Virology*, **118**, 380–388, (1982)

Yoshinaka, Y., Katoh, I. and Luftig, R.B. *Virology*, **136**, 274–281, (1984)

Yoshinaka, Y., Katoh, I., Copeland, T.D. and Oroszlan, S. *Proc. Natl. Acad. Sci. USA*, **82**, 1618–1622, (1985a)

Yoshinaka, Y., Katoh, I., Copeland, T.D., Smythers, G.W. and Oroszlan, S. *J. Virol.*, **57**, 826–832, (1985b)

5
HIV *pol* Expression via a Ribosomal Frameshift

Alan J. Kingsman, Wilma Wilson and Susan M. Kingsman

The genetic relationship of the *gag* and *pol* genes of all retroviruses is approximately the same and the configuration of the protein products of these genes is also strongly conserved (Figure 1) (Weiss *et al.*, 1982).

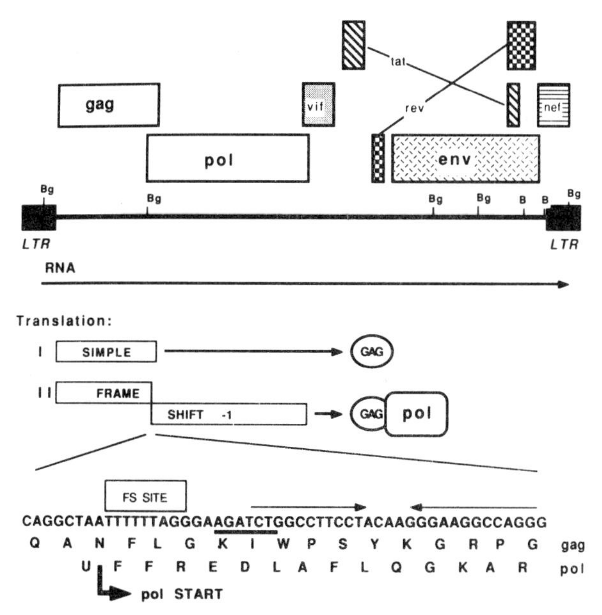

Figure 1.
Expression of *gag* and *pol* in HIV. FS = frameshift. Inverted arrows above the sequence mark the position of a possible stem-loop structure (see text).

The *gag* and *pol* genes are adjacent at the beginning of retroviral transcriptional unit and in most cases the 3' end of *gag* and the 5' end of *pol* overlap by up to a few hundred nucleotides with *pol* in the -1 translational phase with respect to *gag*. Both genes are translated from the full length genomic RNA to produce two primary translation products, a *gag* precursor protein and a GAG:POL fusion precursor protein. The production of the fusion protein requires that the *gag* and *pol* reading frames be brought into translational phase: that is an adjustment of the ribosome's reading phase by a -1 shift is necessary. For several years it was assumed that this shift was mediated by a splice and the absence of any evidence for this was explained by proposing that the splice was small and therefore hard to detect (Weiss *et al.*, 1982). In 1985 two pieces of data suggested that the splicing hypothesis was wrong. First, in the retrovirus-like yeast transposon Ty it was shown that frameshifting between the TYA gene, a *gag* analogue, and the TYB gene, a *pol* analogue, was not due to splicing (Mellor *et al.*, 1985; Clare and Farabaugh, 1985). Secondly, Jacks and Varmus (1985) showed that RSV frameshifting could be achieved in an *in vitro* translation system suggesting that the frameshift was due to some event at the ribosome. The phenomenon is now refered to as ribosomal frameshifting.

Not all retroviruses fuse the products of *gag* and *pol* by the same mechanism. For example in MLV the *gag* and *pol* genes are adjacent and in phase but separated by a UAG termination codon. Production of a GAG:POL fusion protein is achieved by suppression of termination by a glutamine-tRNA (Yoshinaka *et al.*, 1985).

Clearly in retroviruses with *gag* and *pol* out of phase frameshifting is essential for the expression of the enzyme activities of the virus, protease, reverse transcriptase, RNaseH and integrase. Also the frequency of shifting, generally about 5%, determines the relative levels of GAG and POL proteins in the cell. Furthermore the attachment of the POL proteins to the GAG proteins either via the shift, or by termination suppression, not only achieves the genetic economy commonly seen in viruses but also ensures that the enzyme activities are packaged into the GAG core of the virus.

In HIV-1, *gag* and *pol* overlap by 241 nucleotides with *pol* in the -1 phase with respect to *gag* (Figure 2) (Ratner *et al.*, 1985; Wain-Hobson *et al.*, 1985; Sanchez-Pescador *et al.*, 1985). The *gag* precursor protein, Pr55*gag*, is the primary product of simple translation of the full length genomic viral RNA. The GAG:POL fusion protein, Pr160*gag:pol*, also a precursor, is the product of frameshifted translation of the same full length RNA (Figure 1). The frequency of the shift is about 5% and therefore the relative abundance of the two precursors is about 20:1 respectively, although this has only been determined *in vitro* (Jacks *et al.*, 1988a; Wilson *et al.*, 1988).

The HIV overlap begins almost exactly at the 3' end of the p7 coding region. Protein p6 is encoded entirely by the overlap in the GAG phase and the protease coding sequence in the POL phase overlaps p6 by 12 codons (Figure 2).

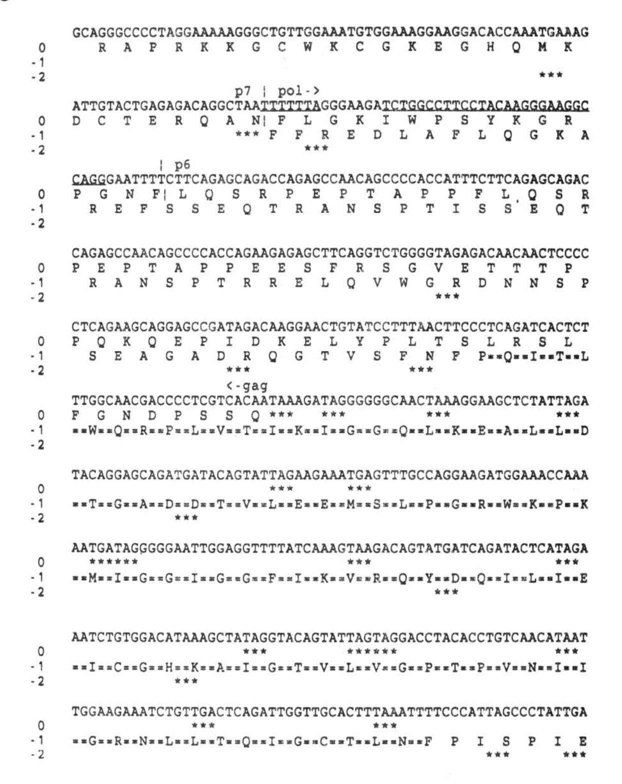

Figure 2.
The nucleotide sequence of the *gag:pol* overlap region of HIV. The sequence is derived from the IIIB isolate reported by Ratner *et al* (1985). The frameshift site and the potential stem– loop region are underlines. 0, –1 and –2 denote translational reading phases where *gag* is arbitrarily given 0. The limits of the *gag* and *pol* open reading frames are marked as are cleavage sites that give rise to the p7 and p6 mature *gag* proteins. The amino acid sequence of the viral protease is marked by '='. Stop codons are marked by '***'.

SEQUENCE REQUIREMENTS FOR THE GAG:POL RIBOSOMAL FRAMESHIFT

The standard assay for frameshifting uses an *in vitro* translation reaction to translate an SP6 generated RNA (Jacks and Varmus, 1985). The RNA is constructed with a preshift sequence, a candidate shift site and then a post-shift sequence (Figure 3). Shifting is detected either by the presence of a second higher molecular weight band in an immunoprecipitation with an antibody directed against the protein encoded by the preshift sequence or by expression of a reporter gene that is dependent on the shift. This system assumes that shifting *in vitro* is the same as shifting *in vivo*. The results that we describe were all obtained with this type of assay. The only exception being the study by Wilson *et al* (1988) where the requirements for HIV frameshifting were investigated in both an *in vitro* translation system and an *in vivo* yeast system.

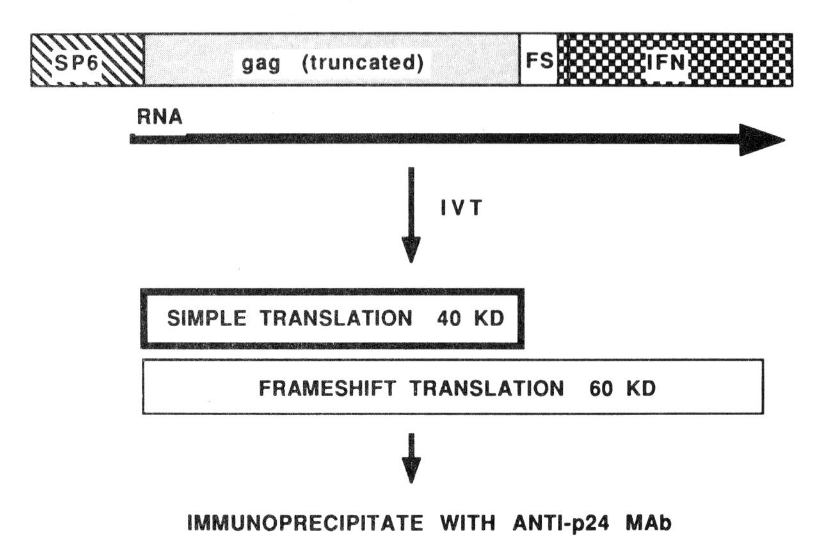

Figure 3.
An SP6 frameshift assay system. IVT = *in vitro* translation. In this case the truncated *nngag* is a pre-shift sequence, FS marks the position of a candidate shifting sequence and IFN (interferon) is a post-shift sequence that adds about 20 kD onto the 40kd GAG protein. This system is taken from Wilson *et al.* (1988).

Frameshifting must occur in the *gag:pol* overlap region so that the shift into the -1, *pol* phase is achieved before the ribosome reaches the stop codon of the *gag* open reading frame. Jacks *et al.* (1988a) showed that shifting occured at codon 3 of the *pol* open reading frame with the sequence TTT.TTA.GGG being read as PHE.LEU.GLY in GAG and as PHE.LEU.ARG in the GAG:POL fusion. However, the data reveal that there is also substantial (at least 30%) phenylalanine in the second position of the shifted product giving an alternative shift site sequence of PHE.PHE.ARG (see later). These data were in agreement with the observation of Wilson *et al.* (1988) who showed that the sequences required for HIV frameshifting were located within the first 16 nucleotides of the overlap region.

The observation that such a short stretch of the overlap region was required for shifting was surprising in the context of what was known about other retroviral requirements for frameshifting (Jacks *et al.*, 1987; Jacks *et al.*, 1988b). In almost every case where a virus makes use of frameshifting as a gene expression strategy the shift site or putative shift site is followed within a very short distance, usually less that 10 nucleotides, by a region of secondary structure (Table 1 and Figure 4)).

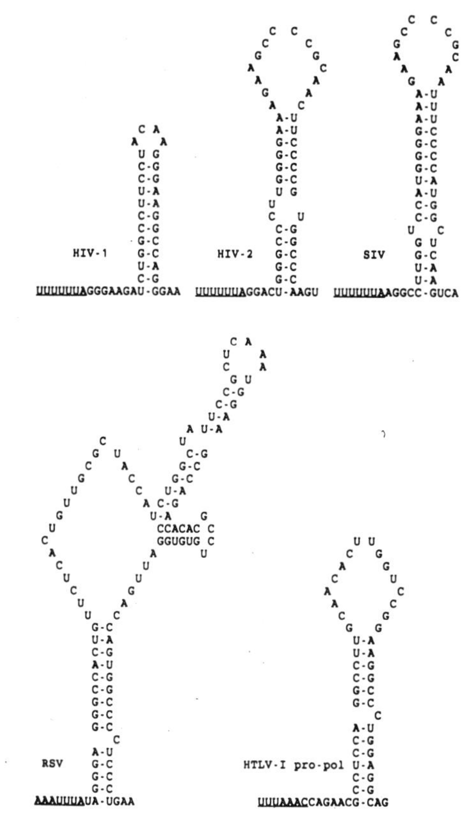

Figure 4.
A selection of potential secondary structures found associated with retroviral frameshift sites. In each case the heptanucleotide shift site is underlined. Sequence information is the same as for Table 1.

The secondary structure may be a simple stem-loop or a pseudoknot in which the loop sequence is capable of base-pairing with another region downstream to form a complex structure (Jacks *et al.*, 1988b; Brierley *et al.*, 1989). Stem-loop structures that may form downstream of a selection of retroviral shift sites, including those of the HIV/SIV family are shown in Figure 4. It is possible that these secondary structures are recognised by soluble 'shifting factors' or by ribosomal components that mediate the shift. A simpler suggestion is that these structures slow the ribosome making it more likely that it will 'slip' back at the shift site. In RSV there is good evidence (Jacks *et al.*, 1988b) that the downstream secondary structure is required from mutational studies that show that destabilisation of the stem-loop substantially decreases frameshifting efficiency and unpublished data (cited in Jacks *et al.* 1988b) indicates that ribosomal pausing may occur at the stem loop. In HIV, however, although a stem-loop structure exists (Figure 4) it does not lie within the first 16 nucleotides of the overlap region. It would seem therefore that HIV frameshifting does not require ribosomal stalling and is therefore not mediated by the same mechanism as is used in RSV (Wilson *et al.*, 1988). We will return to this later.

Shift sites are generally regarded as being heptanucleotide sequences. Examination of these sequences reveals that they fall into two broad groups (Table 1). In what we will call Class I shift sites the first six nucleotides are of the general sequence X.XXY.YY where the stops represent GAG codons. In Class II shift sites the general sequence is X.XXX.XX (Table 1). HIV has a heptanucleotide sequence of U.UUU. UUA. Jacks *et al* (1988a) have shown that mutating the terminal UUUA to CUUA or UCUA dramatically reduces frameshifting efficiency. On the basis of this and the fact that UUUA appears in many shifty sequences it has been suggested that this tetranucleotide is a key signal in determining shifting. However, Wilson *et al* (1988) have shown that deletion of three of the six T/Us in the heptanucleotide also disrupts shifting even though reading frame and the UUUA sequence are preserved. More subtle changes in the 6T/U sequence also disrupt shifting. For example mutation of the third T/U or the sixth T/U to any of the other three nucleotides substantially reduces the shift. However, mutation of the A at position 7 has no effect except in the case of a change of A to T/U when frameshifting increases. This, and other studies (Wilson *et al.*, unpublished data), suggest that the signal for shifting in HIV is the 6T/U sequence rather than UUUA.

The position of the shift means that the GAG:POL fusion protein encodes the GAG mature proteins p17, p24 and p7 but not p6 (Figure 2). The p7 protein contains the RNA binding motif BASIC PATCH-CXXCXXXGHXXXXC. Therefore both GAG and GAG:POL precursor proteins are able to interact with RNA in an immature HIV particle. This may be significant for particle assembly and organisation (see Chapter X). It is also worth noting that the shift site is in a region of the genome that is probably irrelevant as far as protein coding capacity is concerned. This means that co-evolution of a protein sequence and a frameshift is not required.

THE MECHANISM OF RIBOSOMAL FRAMESHIFTING IN RETROVIRUSES

Most studies of viral frameshifting have been carried out on viruses with sequences of the type that we have called Class I, such as RSV (Jacks *et al.*, 1988b) and IBV (Brierley *et al.*, 1989). In both of these, downstream secondary structure, in the form of a stem-loop or a pseudoknot, is required for efficient shifting. The only virus with a Class II sequence that has been analysed is HIV and this does not require a region of secondary structure just downstream of the shift site, even though one exists (Wilson *et al.*, 1988; Madhani *et al.*, 1988). It is tempting, and perhaps useful, to look for an explanation for this in the potential shiftiness of the shift sites alone. In Class I, such as the RSV *gag:pol* frameshift site, there are two short, adjacent, homopolymeric runs of three nucleotides of the general structure X XXY YY (Table 1).

Table 1.
Two classes of retroviral and retroelement frameshifting Examples of demonstrated or suspected shift sites are listed. Sequences are grouped to show *gag* or *gag*-equivalent codons. The last column shows distance from the actual or putative frameshift sequence downstream secondary structure. Rfences for nucleotide sequences: RSV, Rous sarcoma virus (Schwartz *et al.*, 1983); SRV-1, simian retrovirus type 1 (Power *et al.*, 1986); MPMV, Mason-Pfizer monkey virus (Sonigo *et al.*, 1986); 17.6 (Saigo *et al.*, 1984); Visna virus (Sonigo *et al.*, 1985); Mouse IAP, mouse intracisternal A particle (Meitz *et al.*, 1987); BLV, bovine leukemia virus (Sagata *et al.*., 1985; Rice *et al.*, 1985); HTLV-1, human T cell leukemia virus type 1 (Hiramatsu *et al.*, 1987); HTLV-2, human T cell leukemia virus type 2 (Shimotohno *et al.*, 1985; Mador, *et al.*, 1989); HIV-1 (Ratner *et al.*, 1985); HIV-2 (Guyader *et al.*., 1987); SIV, simian immunodeficiency virus (Chakrabarti *et al.*., 1987); gypsy (Marlor *et al.*, 1986); MMTV, mouse mammary tumor virus (Jacks *et al.*, 1987; Moore *et al.*, 1987); EIAV, equine infectious anemia virus (Stephens *et al.*, 1986); Ty1-15 (Mellor *et al.*, 1985; Wilson *et al.*, 1986; Clare, *et al.*, 1988).

Retrovirus/ Retroelement	Overlap	Frameshift Sequence	Distance to SS.
Class I		X XXY YY	
		A AAU UU	
RSV	gag/pol	ACA AAU UUA UAG	7
SRV-1	pro/pol	GGA AAU UUU UAA	8
MPMV	pro/pol	GGA AAU UUU UAA	8
		G GGU UU	
Mouse IAP	gag/pol	CUG GGU UUU CCU	6
		G GGA AA	
SRV-1	gag/pro	CAG GGA AAC GAC	8
MPMV	gag/pro	CAG GGA AAC GGG	8
Visna	gag/pol	CAG GGA AAC AAC	7
		U UUA AA	
BLV	pro/pol	CCU UUA AAC UAG	7
HTLV-1	pro/pol	CCU UUA AAC CAG	7
HTLV-2	pro/pol	CCU UUA AAC CUG	7
Class II		X XXX XX	
		U UUU UU	
HIV-1	gag/pol	AAU UUU UUA GGG	8
HIV-2	gag/pol	GGU UUU UUA GGA	5
SIV	gag/pol	GGU UUU UUA GGC	4
gypsy	gag/pol	AAU UUU UUA GGG	8
		A AAA AA	
MMTV	gag/pro	UCA AAA AAC UUG	8
BLV	gag/pro	UCA AAA AAC UAA	8
HTLV-1	gag/pro	CCA AAA AAC UCC	7
HTLV-2	gag/pro	GGA AAA AAC UCC	8
EIAV	gag/pol	CCA AAA AAC GGG	10
Exception			
MMTV	gag/pol	CAG GUA UUA UGA	5
Ty1-15	TYA/TYB	CAU CUU AGG CCA GAA	

The phase relationship of these two triplets to the *gag* and *pol* reading frames is always the same irrespective of sequence composition. In RSV the heptanucleotide sequence AAAUUUA is the shift site and is thought to mediate shifting through a -1 slip of codon:anticodon interactions at both the A and P sites (Figure 5, Class I) (Jacks *et al.*, 1988b). Following the slip the $tRNA^{leu}$ (UUA) and the $tRNA^{asn}$ (AAU) would be held on the RNA by 2 out of 3 base pairs each. Normal translocation would then take place and translation would proceed, in phase with *pol*, to the end of *pol*. Such a mechanism can be brought about by any adjacent, homopolymeric triplets as long as the distribution of the triplets with respect to the *gag* and *pol* reading phases is the same as in RSV (Table 1). This mechanism does not allow a shift to the +1/-2 phase. In Class II, such as the HIV *gag:pol* frameshift site, there is a single long homopolymeric run of six nucleotides of the general structure X XXX XX (Table 1). Like Class I the relationship of these six nucleotides to the *gag* and *pol* reading phases is conserved and is the same as the relationship of the two triplets in Class I (Table 1). In HIV the heptanucleotide shift sequence is UUUUUUA. Shifting to -1 can be achieved by a mechanism almost identical to that proposed for Class I (Figure 5, Class IIa). In this case the $tRNA^{phe}$ (UUU) and $tRNA^{leu}$ (UUA) would both slip back one nucleotide in the A and P sites. The only difference between Class I and Class II would be that after the slip one of the tRNAs, $tRNA^{phe}$, would be held on the RNA by three out of three base pairs. $tRNA^{leu}$ would be held, as in Class I, by a 'two-out-of- three' association. In Class II, therefore, the slip is maintained by five base pairs as opposed to four in Class I. Consequently, the slip may be more stable. The amino acid sequence over the frameshift region that would be predicted by this mechanism is Asn- Phe-Leu-Arg (Figure 5), the sequence determined for this region by Jacks *et al.* (1988a).

Slippage at the A and P sites is not the only mechanism open to a Class II shift sequence. It is possible that in HIV, for example, the $tRNA^{phe}$ (UUU) slips during translocation (Figure 5, Class IIb) prior to the $tRNA^{leu}$ (UUA) entering the A site. In this case the $tRNA^{phe}$ slips to -1, maintaining three out of three base pairing, and exposing a free UUU codon in the A site. Rather than $tRNA^{leu}$ (UUA) entering the A site, a second $tRNA^{phe}$ (UUU) enters and the -1 shift is completed. The slip is maintained by six out of six base pairs and is likely, therefore, to be quite stable. This mechanism predicts that the amino acid sequence over the HIV frameshift region would be Asn-Phe-Phe-Arg. Both A and P site shifting and translocation shifting seem equally plausible for HIV. If both occured then a mixture of two *gag:pol* products would be produced differing at one amino acid position at the frameshift site. Close examination of the sequence data of Jacks *et al* (1988a) shows exactly that. At least 30% of the HIV-1 shifts produced the sequence Asn-Phe-Phe-Arg. We would suggest, therefore, that the frameshift in HIV might produce microheterogeneity in the *gag:pol* fusion protein.

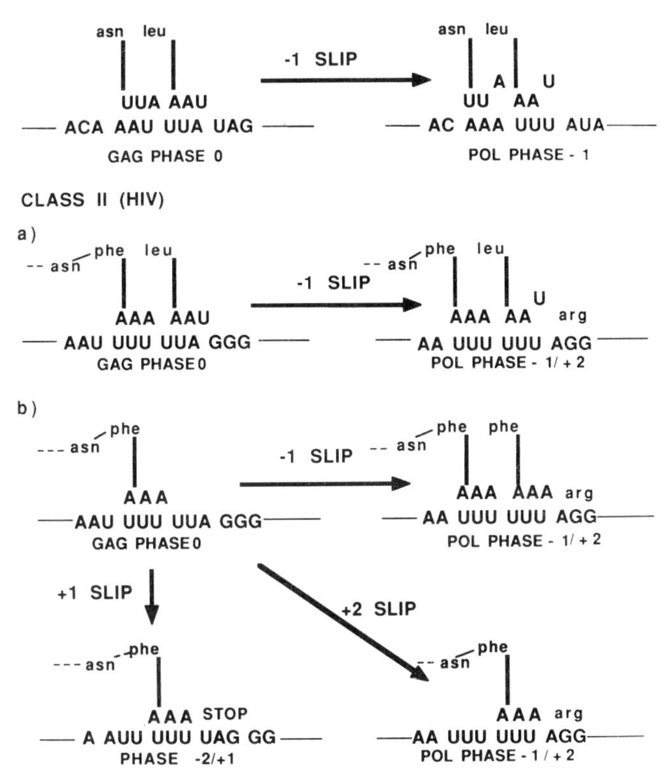

Figure 5.
The mechanisms of Class I and Class II shifting. (see text for explanation).

Translocation slippage of the tRNA^{phe} could also produce a +2 slip (Figure 5, Class IIb) resulting in the amino acid sequence Asn-Phe-Arg. This apparently is not seen (Jacks *et al.*, 1988a). Furthermore the HIV frameshift sequence could mediate shifting into the +1/-2 phase as well as into the -1/+2 phase. The translocation slip of tRNA^{phe} (UUU) could accommodate this via a +1 slip (Figure 5, Class IIb). As there is a UAG termination codon in the +1/-2 phase immediately after the shift site in HIV this observation predicts that there should be a truncated GAG precursor protein of about 48 kD present at about 5% of the level of the authentic p55 GAG precursor protein. Clearly this is testable.

Our scheme suggests that Class II sequences are much shiftier than Class I and this may be reflected in differences in their respective requirements for downstream secondary structures. The requirements for HIV frameshifting appear to be simple. The virus makes use of a generally shifty sequence, T/U6, to express its *pol* gene and this is all that is necessary. In contrast, RSV not only requires a shifty sequence, A3T/U3, but also a region of RNA secondary structure immediately downstream of the shift site (Jacks *et al.*, 1988b). We would suggest that this is not a fundamental difference. Jacks *et al* (1988b) have shown that in RSV the sequence of the region that forms the secondary structure is not critical to shifting efficiency as long as the potential for secondary structure is maintained. This argues against sequence specific cellular factors being involved in a complex mechanism for RSV frameshifting. Class I frameshift sites are likely to be less efficient than Class II sites. There are fewer opportunities for shifting and the slip may be less stable in Class I sites. Viruses with class I, low efficiency, shifty sites may have therefore evolved to use a downstream secondary and possibly tertiary RNA structure (Jacks *et al.*, 1988b; Brierley *et al.*, 1989) to act as a ribosomal stalling site. During the stall there is an increased probability of a shift at the low efficiency site. In contrast, viruses with a class II, high efficiency site do not require a ribosomal stalling site to achieve shifting at the frequency required to deliver the correct ratio of GAG and GAG:POL fusion.

The notion that there are two classes of retroviral and retroelement frameshift sites is easily tested but it also needs to be qualified. There are shift sites that do not conform to either Class I or Class II. The MMTV-1 *pro: pol* shift site is one and may represent a member of a third class. The retrotransposon, Ty, also does not fit into the scheme. Its TYA and TYB gene products are fused by frameshifting occuring in the sequence, CATCTTAGGCCAGAA (in Ty 1-15), (Wilson *et al.*, 1986; Clare, Belcourt and Farabaugh, 1988). This mediates a +1 shift and may, therefore, resemble the +1 shifts observed in some prokaryotic systems, such as the E.coli RF2 gene (Craigen and Caskey, 1986; Curran and Yarus, 1988).

The number of shift sites that have been analysed is remarkably small, Ty, RSV, MMTV, IBV and HIV. A broader study will reveal whether the division of shifting into these two Classes is appropriate or whether the story is more complex.

FRAMESHIFTING *IN VITRO* AND *IN VIVO*

In considering mechanisms of retroviral frameshifting it is important to note that the discussion so far rests on data obtained in *in vitro* translation reactions with SP6 RNA, although one study (Wilson *et al.*, 1988) attempted to move the analysis of HIV frameshifting to an *in vivo* assay by using an efficient expression system in yeast. Conclusions from these systems must be treated with caution as the analytical procedures used to date are far removed from the *in vivo* situation in an HIV infected human cell. In particular the role of secondary structure in determining shifting in HIV needs to be tested *in vivo*. There is no doubt that in the *in vitro* and yeast systems used so far there is no requirement for downstream secondary structure, but one cannot ignore the fact that such structures are strongly conserved in HIV and SIV (Figure 4). These considerations become even more important as the peculiarities of RNA handling by HIV emerges.

There are at least two proteins, TAT and REV, encoded by the virus that mediate regulation of viral expression (Cullen and Green, 1989). They appear to function by acting on viral RNA . REV controls the relative expression capabilities of spliced and unspliced RNAs by a process that appears to involve the selective transport of unspliced RNA to the cytoplasm (Malim *et al.*, 1989). This is probably brought about by the recognition of the REV Response Element (RRE) on the RNA by REV itself or a complex containing REV. Intriguingly, REV must localise to the nucleolus for this apparent transport to take place. While the mechanism of REV action is still unclear it is apparent that the 'REV pathway' is required for expression of the unspliced full length HIV RNA that encodes GAG and GAG:POL. The mode of action of TAT is even less clear. It acts to potentiate the expression of all viral RNAs by both transcription (e.g. Laspia *et al.*, 1989) and post- transcription events (Braddock *et al.*, 1989). This potentiation is mediated via a TAT Activation Region (TAR) at the 5' end of viral RNAs (Braddock *et al.*, 1989; Berkhout *et al.*, 1989). Both the transcriptional and post-transcriptional activities of TAT require localisation of both TAT and its target RNA in the nucleus (Braddock *et al.*, 1989). It seems likely, though it is by no means proven, that the 'TAT pathway' links transcription and post- transcriptional events in some way and represents some form of specialised RNA handling system.

In considering HIV frameshifting in the context of the virus therefore it is important to remember that the RNA that is subject to shifting may be shifting in the context of complex RNA handling pathways mediated by TAT and REV. This might include packaging of the RNA into complex RNPs and translation in unconventional cellular environments. In these circumstances the determinants for frameshifting may be quite different from those that function *in vitro* or in the absence of the RNA handling pathways *in vivo*, such as in yeast.

ACKNOWLEDGEMENTS

We would like to thank our colleagues in the Virus Molecular Biology Group for comments on the manuscript. The MRC supports our work on HIV frameshifting.

REFERENCES

Berkhout, B., Silverman, R.H. and Jeang, K-T. *Cell*, **59**, 273-282, (1989).

Braddock, m., Chambers, A., Wilson, W., Esnouf, M.P., Adams, S.E., Kingsman, A.J. and Kingsman, S.M. *Cell*, **58**, 269-279 (1989)..

Brierley,I., Digard,P. and Inglis,S.C. *Cell*, **57**, 537-547,(1989).

Chakrabarti,L., Guyader,M., Alizon,M., Daniel,M.D., Desrosiers,R.C., Tiollais,P. and Sonigo,P. *Nature*, **328**, 543-547, (1987).

Clare, J. and Farabaugh, P. *Proc. Natl. Acad. Sci. USA*, **82**, 2829-2833, (1985).

Clare,J.J., Belcourt,M. and Farabaugh,P.J. *Proc. Natl. Acad. Sci. USA*, **85**, 6816-6820, (1988).

Craigen,W.J. and Caskey,C.T. *Nature*, **332**, 273-275, (1986).

Cullen, B.R. and Greene, W.C. *Cell*, **58**, 423–426, (1989).
Curran,J.F. and Yarus,M. *J. Mol. Biol.*, 203, 75–83, (1988).
Guyader,,M., Emerman,M., Sonigo,P., Claver,F., Montagnier,L. and Alizon,M. *Nature*, **326**, 662–669, (1987).
Hiramatsu,K., Nishida,J., Naito,A. and Koshikura,H. *J.Gen. Virol.*, **68**, 213–218, (1987).
Jacks,T. and Varmus,H.E. *Science*, **230**, 1237– 1242, (1985).
Jacks,T., Townsley,K., Varmus,H.E. and Majors,J. *Proc. Natl. Acad. Sci. USA*, **84**,4398–4302, (1987).
Jacks,T., Power,M.D., Masiarz,F.R., Luciw,P.A., Barr,P.J. and Varmus,H.E. *Nature*, **331**, 280–283, (1988a).
Jacks,T., Madhani,H.D., Masiarz,F.R. and Varmus,H.E. *Cell*, **55**, 447–458, (1988b).
Laspia, M.F., Rice, A.P. and Mathews, M.B. *Cell*, **59**, 283–292, (1989).
Madhani,H.D., Jacks,T. and Varmus,H.E. in 'The Control of HIV *pol* Gene Expression', R.Franza, B.Cullen and F.Wong-Staal,eds. (Cold Spring Harbor, New York: Cold Spring Harbor Laboratory), pp. 119–125, (1988).
Mador,N., Panet,A and Honigman,A. *J.Virol.*, **63**, 2400–2404, (1989).
Malim, M.H., Hauber, J., Le, S-Y., Maizel, J.V. and Cullen, B.R. *Nature*, **338**, 254–257, (1989).
Marlor,R.L. Parkhurst,S.M. and Corces,V.G. (1986). *Mol. Cell. Biol.*, **6**, 1129–1134.
Meitz,J.A., Grossman,Z., Leuders,K.K. and Kuff,E.L. *J. Virol.*, **61**, 3020–3029, (1987).
Mellor,J., Fulton,A.M., Dobson,M.J., Wilson,W., Kingsman,S.M. and Kingsman,A.J. *Nature*, **313**, 243–246, (1985).
Moore,R., Dixon,M. Smith,R., Peters,G. and Dickson,C. *J. Virol.*, **61**, 480–490, (1987).
Power,M.D. Marx,P.A. Bryant,M.L., Gardner,M.D., Barr,P.J. and Luciw,P.A. *Science*, **231**, 1567– 1572, (1986).
Ratner,L., Haseltine,W., Patarca,R., Livak,K.J., Starcich,B., Josephs,S.F., Doran,E.R., Rafalski,J.A., Whitehorn,E.A., Baumeister,K., Ivanoff,L., Petteway,Jr.,S.R. Pearson,M.L., Lautenberger,J.A., Papas,T.S., Ghrayeb,J., Chang,N.T., Gallo,R.C. and Wong-Staal,F. *Nature*, **313**, 277–284, (1985).
Rice,C.M. and Strauss,J.H. *Proc. Natl. Acad. Sci. USA*, **78**, 1062–1066, (1981).
Sagata,N., Yasunaga,T., Tsuzuku-Kawamura,J., Ohishi,K., Ogawa,Y. and Ikawa,Y. *Proc. Natl. Acad. Sci. USA*, **82**, 677–681, (1985).
Saigo,K., Kugimiya,W., Matsuo,Y., Inouye,S., Koshioka,K. and Yuki,S. *Nature*, **312**, 659–661, (1984).
Sanchez-Pescador,R., Power,M.D., Barr,P.J., Steimer, K.S., Stempien,M.M., Brown-Shimmer,S.L., Gee,W.W., Renard,A., Randolph,A., Levy,J.A., Dina,D. and Luciw,P.A. *Science*, **227**,484–492, (1985).
Schwartz,D.E., Tizard,R. and Gilbert,W. *Cell*, **32**, 853–869, (1983).
Shimotohno,K., Takahashi,Y., Shimizu,N., Gojobori,T., Golde,D.W., Chen,I.S.Y., Miwa,M. and Sugimura,T. *Proc. Natl. Acad. Sci. USA*, **82**, 3101–3105, (1985).
Sonigo,P., Alizon,M., Staskus,K., Klatzmann,D., Cole,S., Danos,O., Retzel,E., Tiollais,P., Haase,A. and Wain-Hobson,S. *Cell*, **42**, 369–382, (1985).
Sonigo,P., Barker,C., Hunter,E. and Wain-Hobson,S. *Cell*, **45**, 375–385, (1986).

Stephens,R,M., Casey,J.W. and Rice,N.R. *Science*, **231**, 589–594, (1986).
Wain-Hobson,S., Sonigo,P., Danos,O., Cole,S. and Alizon,M. *Cell*, **40**, 9–17, (1985).
Weiss,R., Teich,N., Varmus,H. and Coffin,J. 'Molecular Biology of Tumor Viruses.' (Cold Spring Harbor, New York: Cold Spring Harbor Laboratory), (1982).
Wilson,W., Malim,M.H., Mellor,J., Kingsman,A.J. and Kingsman,S.M. *Nucl. Acids Res.*, **14**, 7001–7015, (1986).
Wilson,W., Braddock,M., Adams,S.E., Rathjen,P.D., Kingsman,S.M. and Kingsman,A.J. *Cell*, **55**, 1159–1169, (1988).
Yoshinaka, Y., Katoh, I., Copeland, T.D. and Oroszlan, S. *Proc. Natl. Acad. Sci. USA* **82**, 1618–1622, (1985).

6
Functional Characterisation of HIV-1 *gag-pol* Fusion Protein

Cheng Peng, Karin Moelling, Nancy T. Chang and Tse Wen Chang

Among the three major genes, *gag*, *pol*, and *env* of the human immunodeficiency virus (HIV), the *pol* gene encodes at least three products, protease, reverse transcriptase (RT), and integrase, which are essential for viral replication and maturation (Steimer *et al*, 1986; Lillehoj *et al*, 1988 ; Kohl *et al*, 1988; Peng *et al*, 1989). The expression of functional *pol* gene products has been proposed to include two important stages. The first is that by means of a ribosomal frameshifting mechanism *gag-pol* fusion protein, which contains a large portion of *gag* coding region and the entire *pol* coding region is generated. Published *in vitro* transcription and translation studies have shown that the *gag* and *pol* genes of HIV-1 overlap and use two different translational reading frames (Jacks *et al*, 1988). These studies have also shown that the *gag-pol* protein is created by a ribosomal '-1 frameshifting' mechanism during translation, resulting in a long, fused polypeptide composed of a C-terminal-truncated *gag* peptide and the entire *pol* peptide. The second stage involves the proteolytic processing of this fusion protein to generate mature structural and functional viral components (Debouck *et al*, 1987; Lillehoj *et al*, 1988). An HIV-specific protease cleaves the fusion protein to generate matured core proteins, RT, integrase, and protease.

According to this model, several vital viral products are derived from the common *gag-pol* fusion protein precursor. It is, therefore, important to establish the existence of this polyprotein precursor and characterize its properties. However, under natural conditions, the putative precursor presum ably undergoes relatively rapid processing and would be very difficult to isolate and to characterize (Hansen *et al*, 1988; Mous *et al*, 1988). So far, for HIV, there have been limited studies that convincingly establish the existence of this precursor protein either in cells or in viral particles. (Starnes *et al*, 1988; Lori *et al*, 1988) Recently, several investigators (Leuthardt and Le Grice, 1988; Le Grice *et al*, 1988; Tanese *et al*, 1988) studied the function of *gag-pol* protein using an *E. coli* expression system. They described that *E. coli*-produced *gag-pol* fusion protein contained no RT activity and suggested that the RT had to be processed to the end products (66 and 55 Kda) to acquire enzymatic activity. In this report, we present the studies using mammalian cell expression system to analyze *gag-pol* product, its processing, and its enzymatic and antigenic activities. We have shown that the protease-defective HIV particles contained RT activity and that this RT activity resided on the unprocessed *gag-pol* fusion polypeptide.

RESULT AND DISCUSSION

Construction of Mutants

The HIV protease was assumed to be the major proteolytic enzyme causing the breakdown of *gag-pol* polyprotein precursor. A panel of protease defective HIV mutants, which we constructed for studying the role of the protease in viral infectivity, were also used in the present studies (Fig. 1). Two deletion mutants, AH2 and BH27, which were derived from the wild-type proviral genome, HXB2, were selected (Ratner *et al*, 1987; Peng *et al*, 1989). Mutant AH2, which lost 39 nucleotides in the protease coding region and acquired 12 nucleotides from *Bam* HI linker maintained the same reading frame as the wild-type. In mutant BH27, which lost 71 nucleotides in the protease coding region and also acquired the 12 nucleotides linker, a new translational termination codon was created two amino acids downstream from the mutation site. Considering that a deletion mutation may affect the protein structure to a greater degree than a point mutation, we also constructed an HIV variant with a point mutation in a gene segment corresponding to the putative active enzymatic site of the protease (Pearl and Taylor, 1987). The genome of this site-specific mutant was produced by polymerase chain reaction (PCR) and contained a single nucleotide substitution, A to C (encoding amino acid residue, Asp to Ala). This site-specific mutant had the identical amino acid residue replacement as the one employed by Le Grice *et al*, (1988). All of the wild-type and mutated viral genomes were cloned to a pSVL vector, which has an SV40 late promoter in front. We denoted those constructed plasmids as HXB2-pSVL for the wild-type, AH2-pSVL for the in-frame deletion mutant, BH27-pSVL for the out-frame deletion mutant, and AC-pSVL for the site-specific mutant.

Expression of wild-type and mutant viruses.

The constructed wild-type and mutated viral genomes were used to transfect the SV40-transformed monkey kidney cell line COS-M6 with DEAE-dextran (Lopata *et al*, 1984). The culture supernatant of the transfected COS-M6 cells were examined with electron microscope to determine whether viral particles existed or only soluble sub-virion components were expressed. All of the wild-type and mutant virions were budded and released into the extracellular space. Most of the wild-type virions had envelopes, central cores and eccentric nucleoids morphologically characteristic of HIV particles produced by infected lymphocytes. The protease-defective virions were frequently budded into cytoplasmic vacuoles and lacked electron-dense (cylindrical) cores but contained ring-shaped nucleoids. Vector pSVL transfected cells did not produce viral particles (Peng *et al*, 1989).

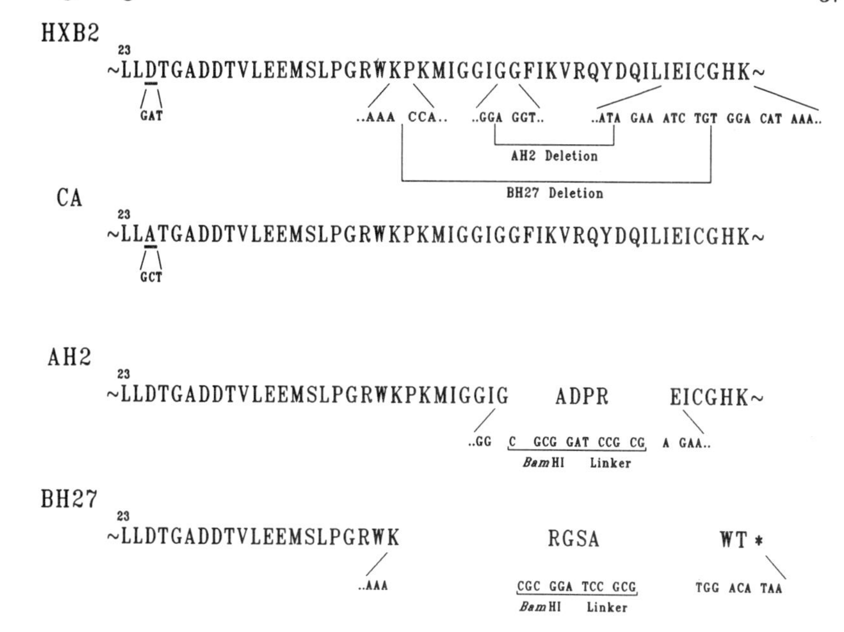

Figure 1
Amino acid and nucleotide sequences around the mutation sites of constructured HIV-1 mutants. HXB2 is the wild-type proviral genome. The first amino acid residue in these sequences is the 23rd amino acid residue of mature HIV protease. CA is the site-specific mutant with a single nucleotide substitution and the amino acid residue involved in this point mutation underlined. AH2 and BH27 mutants were made by removing nucleotids from HXB2 with *Bal* 31 nuclease at restriction enzyme *Bcl* I site. '*' represents a termination codon.

The virions synthesized by COS-M6 cell were also examined for their infectivity. Human T-cell lymphoma cell lines, H9 and MT-4, were incubated with serially diluted culture supernatants of COS-M6 cells transfected with the various HIV genomes. Both antigen capture assay and reverse transcriptase assay demonstrated that the protease-defective HIV did not cause productive infection (Kohl *et al*, 1988; Peng *et al*, 1989; Fig. 2).

While wild-type and mutated viruses could be produced by COS-M6 cells, it is clear that no cellular protease from this cell line could process the *gag* and *gag-pol* products (Fig. 2; Peng *et al*, 1989). Since HIV mainly infects lymphoid cells, we also investigated whether cellular proteases existed in lymphoid cells that could digest *gag* products. For these experiments, we prepared an enzyme substrate, a partial *gag* peptide, using an *in vitro* 'run-off' transcription procedure and a cell-free translation system with rabbit reticulocyte lysates. The *gag* peptide was labeled with ^{35}S methionine. In solid-phase (fixed *S. aureus* beads).

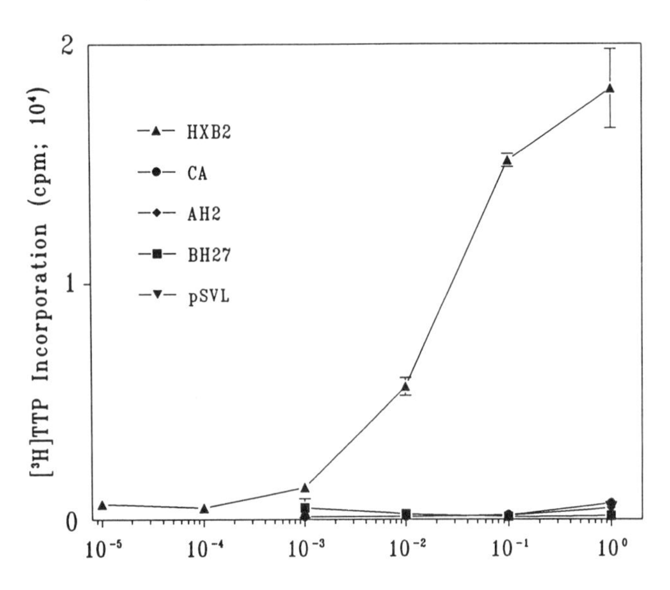

Figure 2
Infectivity of wild-type and mutant HIV-1 on MT-4 cells. The viruses were produced and harvested from the culture supernatants of transfected COS-M6 cells and serially diluted for testing infectivity on MT-4 cells. RT activity in the culture media of the MT-4 cells was determined at day 5. A sample from of COS-M6 cells transfected with vector pSVL was used as the negative control. Each value is mean ± standard deviation of two determinations.

immunoprecipitation with mAb against HIV *gag* p24 (C246), 41 kDa product could be clearly identified. Cell lysates of H9 cells or HTLVIIIB-infected H9 cells were prepared by freezing and thawing three times and were incubated with substrate-antibody-solid phase support complex. The reacted mixture were then analyzed by SDS-PAGE and autoradiography. The results showed that the lysates from HIV-1-infected H9 cells and not from uninfected cells could processed the ^{35}S-labeled *gag* peptide to generate smaller molecules (Fig. 3), suggesting that only HIV-1-encoded protease could process the *gag* precursor to mature core proteins.

Structural Characterization of Viral *gag*, *pol*, and *gag-pol* Products.

Our studies have focussed on two of the viral products derived from the *gag* and *pol* genes and employed monoclonal antibodies (mAbs) specific for p24 (C246) and RT (C14120) to detect and analyze these two viral

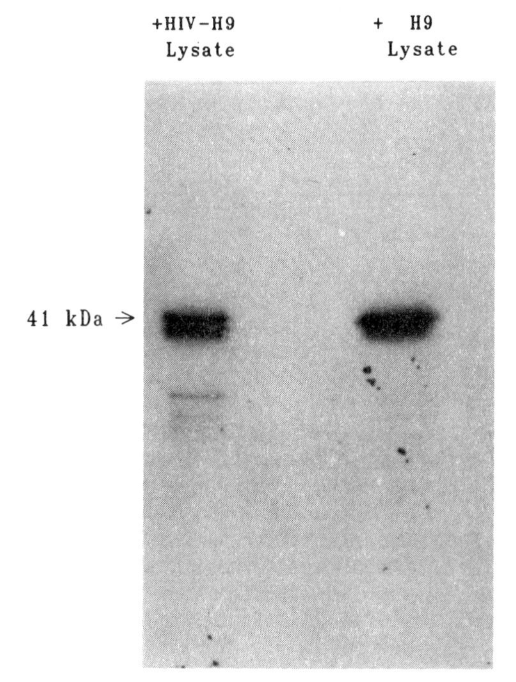

Figure 3
Processing of viral protein by HIV specific protease. ^{35}S-methionine labeled protease substrate was synthesized by *in vitro* transcription and cell-free translation system. Followed by solid-phase immunoprecipitation with anti-p24 mAb, C246, the immune complexes were then incubated with cell lysates from either HIV-infected H9 cells or H9 cells. The reaction mixture were analyzed by SDS-PAGE and autoradiography.

components. Using mAbs specific for p24 and RT as probes in immunoblotting assays, the wild-type viruses were shown to contain processed *gag* p24 protein and *pol* RT of 55 and 66 kDa. The protease-defective mutants, AH2, CA and BH27, contained the precursor of *gag* protein Pr56^{gag}. In addition, both mAb anti-p24 and anti-RT revealed bands of about 160 kDa in mutants AH2 and CA. But in the *pol* region truncated mutant, BH27, only an extra band of 57 kDa was shown to be reactive with mAb to p24.

Although both mAb specific for p24 and for RT were reactive with a protein of the same molecular weight, we attempted to obtain additional evidence to show that these two antibodies were against the same molecule, which was presumably the *gag-pol* fusion product. For this purpose, subtractive immunoprecipitation experiment was performed. Viral particles were lysed with detergent (0.5% NP-40) with a high concentration of salt (0.8 M NaCl). After the salt concentration was adjusted to 0.15 M NaCl, the mAb C246 was added to bind to any viral product with p24 epitope. Solid phase adsorbents combining fixed *S. aureus* beads and the same beads covalently conjugated with goat anti-mouse antibody ('Tachisorb M IgG Immunoadsorbent' from Calbiochem) were used to remove the immune complexes and free antibodies. The adsorbed viral lysates were then subjected to SDS-PAGE, immunoblotting, and detection with mAb against RT. The results showed that when C246 removed molecules of 160 kDa and others with p24 epitope, it also eliminated the 160 kDa molecule with reactivity to anti-RT, indicating clearly that the 160 kDa molecule contained both *gag* and *pol* segments (Peng *et al*, manuscript in preparation). Thus after the existence of HIV *gag-pol* fusion protein has been proposed for years, we provide the direct evidence to demonstrate that this *gag-pol* precursor protein is indeed synthesized in the cells and can also be packaged into the virions.

Functional Characterization of *gag-pol* Products.

As metioned above, our earlier work has also shown that mutant AH2 virions contain RT activity and we believe that this activity was possessed by the unprocessed *gag-pol* protein (Peng *et al*, 1989). In the studies of HIV protease function by mutational analyses, Le Grice and coworkers (Le Grice *et al*, 1988; Leuthardt and Le Grice, 1988) suggested that HIV RT protein must be processed from the polyprotein to acquire its enzymatic activity. One explanation for this discrepancy is that the polypeprotein produced in the *E. coli* expression system used by these investigators did not have proper post-translational modifications as it would have in mammalian cells and, therefore, did not have RT activity. Another possibility exists that the deletion mutation in AH2 genome may alter the overall structure of the *gag-pol* fusion protein and render the RT moiety to acquire some enzymatic activity which is nonexistent in the wild-type precursor protein. We have, therefore, constructed a mutant CA with the identical point mutation in the protease region as constructed by Le Grice *et al* (1988). It was found that the mutant CA lysate has as much RT activity as AH2 lysate (Fig. 4).

We have also performed subtractive immunoadsorption experiments using anti-p24 mAb, C246, as discussed above. Instead of running SDS-PAGE and immunoblotting of viral lysates after the removal of molecules with p24 epitope with mAb C246, the RT activity of the adsorbed lysates was measured. The results showed that the RT activity of both AH2 and CA mutants was decreased by about 40% by the adsorptive treatment with the anti-p24 mAb (Table 1). The RT activity from wild-type HIV showed only very limited reduction. We thus conclude that the *gag-pol* fusion protein does have RT activity. The results also ruled out the possibility that trace amounts of mature RT generated from non-specific cellular protease were packaged in protease-defective mutant virions and contribute to most of the RT activity observed.

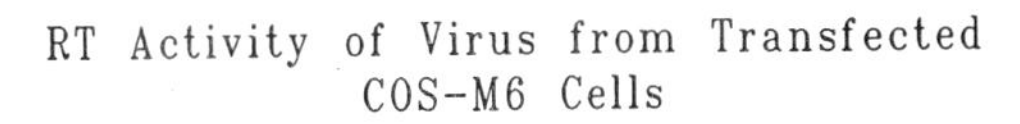

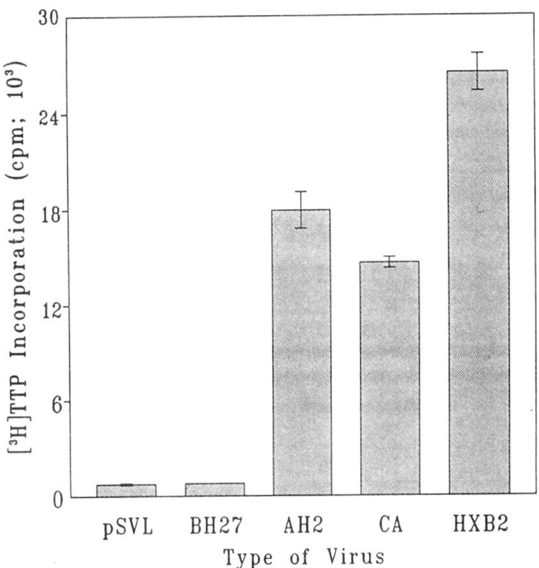

Figure 4
RT activity of wild-type and mutant HIV-1. COS-M6 cells were transfected with constructed plasmids containing HXB2, CA, AH2, BH2 or vector pSVL. The culture supernatants were collected at 72h after transfection. The viral particles were lysed by detergent and RT activity measured by the incorporation of ^{3}H-labeled TTP.

Table I
Subtractive Immunoadsorption Assay of RT Activity.

Type of Virus	RT Activity (cpm) Adsorption with		% of Adsorption
	OKT4	anti-p24	
HXB2	42,614	47,470	-11.4*
CA	11,571	6,737	41.78
AH2	15,783	9,463	40.04
BH27	387	454	-
pSVL	170	203	-

*Percentage of RT adsorpted was calculated as (cpm with OKT4 - cpm with anti-p24)/(cpm with OKT4) * 100%

CONCLUSION

Studying the structure and function of HIV-1 *gag-pol* polyprotein is important because several essential viral components are derived from it. By using HIV-1 protease-defective mutants, we have established the existence of *gag-pol* fusion protein in mammalian cells transfected with HIV-1 genomes and found that it has significant RT activity. We believe that *E. coli*-expressed *gag-pol* fusion protein probably does not possess a proper structure and therefore, does not have RT activity. Presently, we are investigating whether protease-defective HIV-1 mutants, such as AH2 and CA, which contain unprocessed *gag-pol* precursor with some reverse transcriptase activity, can enter target cells and carry out some degree of synthesis of viral DNA.

REFERENCES

Debouck, C., Gorniak,J.G., Strickler,J.E., Meek,T.D., Metcalf,B.W. and Rosenberg, M. *Proc. Natl. Acad. Sci. USA*, **84**, 8903-8906, (1987).

Hansen,J., Billich,S., Schulze,T., Sukrow,S., and Moelling,K. *EMBO* J, **7**, 1785-1791, (1988).

Jacks,T., Power,M.D., Masiarz.F.R., Luciw,P.A., Barr,P.J., and Varmus, H.E. *Nature (London)*, **331**, 280-283, (1988).

Kohl,N,E., Emini,E.A., Schleif,W.A., Davis,L.J., Heimbach,J.C., Dixon,R. A.F., Scolnick,E.M., and Sigal,I.S. *Proc. Natl. Acad. Sci. USA*, **85**, 4686-4690, (1988).

Le Grice,S.F.J., Mills,J., and Mous,J. *EMBO J.*, **7**, 2547-2553, (1988).

Leuthardt,A., and Le Grice,S.F.J. *Gene*, **68**, 35-42, (1988).

Lillehoj,E.P., Salazar,F.H.R., Mervis,R.J., Raum,M.G., Chan,H.W., Ahmad,N., and Venkatesan,S. *J. Virol.*, **62**, 3053-3058, (1988).

Lopata,M.A., Cleveland,D.W., and Sollner-Webb,B. *Nucleic Acids Res.*, **12**, 5707-5717, (1984).

Lori,F., Scovassi,A.I., Zella,D., Achilli,G., Cattaneo,E., Casoli,C,, and Bertazzoni,U. *AIDS Res. Hum. Retroviruses*, **4**, 393-398, (1988).

Mous,J, Heimer,E.P., and Le Grice,S.F.J. *J. Virol.*, **62**, 1433-1436, (1988).

Pearl,L.H., and Taylor,W.R. *Nature (London)*, **329**, 351-354 (1987).

Peng,C., Ho,B.K., Chang,T.W., and Chang,T.N. *J. Virol.*, **63**, 2250-2256, (1989).

Ratner,L., Fisher,A., Jagodzinski,L.L., Mitsuya,H., Liou,R.-S., Gallo,R.C., and Wong-Staal,F. *AIDS Res. Hum. Retroviruses*, **3**, 57-69, (1987).

Starnes,M.C., Gao.,W.Y., Ting,R.Y., and Cheng,Y.C. *J. Biol. Chem.*, **263**, 5132-5134, (1988).

Steimer,K.S., Higgins,K.W., Powers,M.A., Stephans,J.C., Gyenes,A., George-Nascimento,C., Luciw,P.A., Barr,P.J., Hallewell,R.A., and Sanchez-Pescador,R. *J. Virol.*, **58**, 9-16, (1986).

Tanese,N., Prasad,V.R., and Goff,S.P. *DNA*, **7**, 407-416, (1988).

7
Biosynthesis and Processing of the *gag* and *pol* Polyproteins of Human Immunodeficiency Virus Type 1 in *Escherichia coli*

Mei Huei T. Lai, Albert G. Dee, Peter H. Zervos, William F. Heath, Jr
and Maurice E. Scheetz

The three major structural genes of human immunodeficiency virus type 1 (HIV-1) are arranged in the viral genome in the order of 5' *gag-pol-env* 3'. The gag gene encodes four group specific antigens, MA(p17), CA(p24), NC(p7) and p6, while the *pol* gene codes for protease, reverse transcriptase/ribonuclease H (RT/RH, 66 kDa) and integrase (IN, 32 kDa) (Chassagne *et al.*., 1986; Henderson *et al.*, 1988; Kramer *et al.*, 1986; Lightfoot *et al.*, 1986; Lillehoj *et al.*, 1988). The primary translational product of the *gag* gene is a 55 kDa polyprotein (pr55^{gag}) (Kalyanarman *et al.*, 1984; Sarngadharan *et al.*, 1985), and that of the *pol* gene is a *gag-pol* fusion protein 160 kDa in size (pr160$^{gag-pol}$) achieved by ribosomal frameshift near the 3' end of the *gag* gene (Gendelman *et al.*, 1987; Jacks *et al.*, 1988). They are processed to the individual viral proteins by the protease encoded by the *pol* gene after the enzyme cleaves itself from the

HIV-I Protease Cleavage Sites and Surrounding Sequences in the GAG and POL Polyproteins of Clone HXB2 of HTLV$_{IIIB}$ Isolate

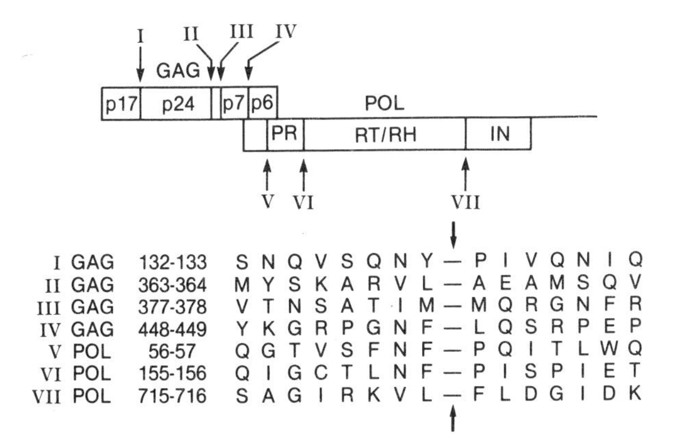

I GAG	132-133	S	N	Q	V	S	Q	N	Y	—	P	I	V	Q	N	I	Q
II GAG	363-364	M	Y	S	K	A	R	V	L	—	A	E	A	M	S	Q	V
III GAG	377-378	V	T	N	S	A	T	I	M	—	M	Q	R	G	N	F	R
IV GAG	448-449	Y	K	G	R	P	G	N	F	—	L	Q	S	R	P	E	P
V POL	56-57	Q	G	T	V	S	F	N	F	—	P	Q	I	T	L	W	Q
VI POL	155-156	Q	I	G	C	T	L	N	F	—	P	I	S	P	I	E	T
VII POL	715-716	S	A	G	I	R	K	V	L	—	F	L	D	G	I	D	K

Figure 1.
HIV-1 protease cleavage sites in the *gag* and *pol* polyproteins.

pr160$^{gag-pol}$ precursor by autocatalysis (Farmerie *et al.*, 1987; Graves *et al.*, 1988). There are at least seven protease cleavage sites in the *gag* and *pol* polyprotein precursors (Darke *et al.*, 1988), as depicted in Figure 1. The numbering of amino acid residues at the cleavage sites is based on the amino acid sequence predicted from the nucleotide sequence of clone HXB2 of HIV-1 isolate, HTLV IIIB (Ratner *et al.*, 1985).

Little is known about the temporal events involved in the biosynthesis and processing of the *gag* and *pol* polyprotein precursors. This is probably due in part to the experimental difficulties in following closely these events in cells infected with HIV-1. To circumvent this problem, we constructed several prokaryotic vectors to express the *gag* and *pol* genes in *E. coli* under the transcriptional control of the P_L promoter and a thermal labile repressor, CI^{857} derived from bacteriophage lambda. The advantage of this expression system is the fact that the repressor CI^{857} is rapidly inactivated by heat. The expression of any gene driven by the P L promoter can be turned on quickly by elevating the growth temperature from 32° C to 40° C. In this report, we describe the biosynthesis and the sequential processing of the *gag* and *pol* polyproteins of HIV-1 in *E. coli* .

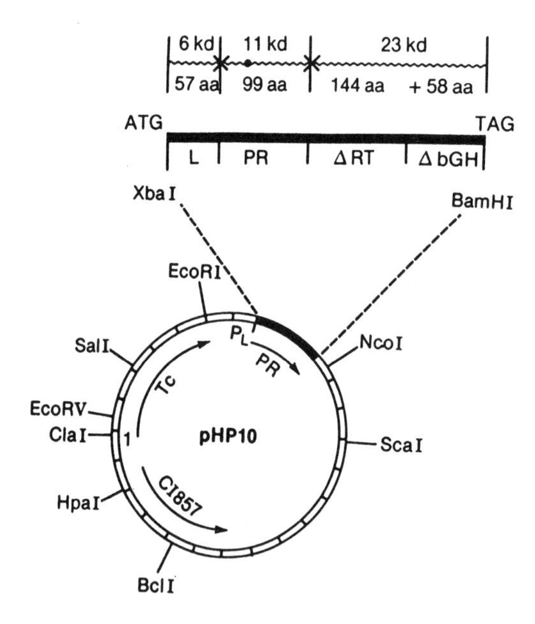

Figure 2.
Plasmid map of pHP10 and organization of proteins encoded by the DNA sequence (marked PR in the plasmid map) inserted downstream and driven by the P_L promoter of bacteriophage lambda regulated by thermal labile repressor CI^{857}.

EXPERIMENTAL RESULTS AND DISCUSSION

The processing of the *pol* polyprotein.

The parental expression vector, pL110c, was constructed for expression of bovine growth hormone (bGH) (H. Hsiung, Lilly Research Laboratories. Details will be published elsewhere). The coding sequence for bGH was inserted between restriction sites *Xba*I and *Bam*HI located downstream of the P_L promoter. Expression vector pHP10 (Figure 2) was constructed to study the autocatalysis of HIV-1 protease produced as a truncated *pol* polyprotein. It contained nucleotides (nt) 2084 to 2978 of HIV-1 clone HXB2. The vector was constructed by ligating a *Bgl* II-*Eco* RV fragment of clone HXB2 (nt 2096-2978) upstream of a synthetic oligonucleotide containing an *Xba* I recognition sequence, a ribosome binding site, and *Nde* I restriction site (CATATG), and the first four codons of the *pol* open reading frame (ORF) that were cleaved off by *Bgl* II digestion. The hybrid molecule was then inserted into *Xba* I and *Sma* I sites in pL110c, resulting in in frame fusion with the last 58 codons of the bGH gene. Initiating at the ATG codon in *Nde* I site, the insert encoded the first 56 amino acids (aa) of the *pol* ORF, the mature protease (99 aa), the first 144 aa of reverse transcriptase, and the last 58 aa of bGH, designated in Figure 2 as L, PR, ΔRT and ΔbGH, respectively.

Plasmid pHP10 was transformed into E. coli strain L507 (htpR165 am, lonR9, cps3, derived by S. R. Jaskunas, Lilly Research Laboratories, from LC137 strain originally provided by F. Goldberg, Harvard University). The transformed cells were grown at 32°C until mid log phase. The cultivation temperature was quickly elevated to 40°C to induce gene expression. At 0, 2.5, 5, 10,15, 30, 45, and 75 minutes after induction, samples were taken and analyzed by 15% SDS polyacrylamide gel electrophoresis and Western blot analysis, using a rabbit polycolonal antiserum raised against a synthetic peptide with sequence corresponding to residues 17-36 of the mature HIV protease. This region includes the active site of the enzyme (Kohl *et al.*, 1988; Pearl and Taylor, 1987). The results are shown in Figure 3. As predicted from the coding capacity of the insert in pHP10, L5-10 cells (L507 cells transformed by pHP10) first synthesized a 40 kDa protease containing protein, which appeared as early as 10 minutes after induction (Figure 3, lane 4). It was then followed by the appearance of a 34 kDa intermediate. The mature protease, migrating as 11 kDa protein, was detected clearly at 30 minutes after induction (Figure 3, lane 6). Its production appeared to peak at 75 minutes.

For protease to cleave itself from the 40 kDa precursor, the autocatalysis could occur via one of the two pathways illustrated in Figure 4. If the protease autoprocessed at its N terminus first, a protease containing intermediate 301 aa long (~33-34 kDa) should be generated, followed by the appearance of the 11 kDa mature protease (pathway A). If the protease cleaved itself at the C terminus first, a protease containing intermediated 156 aa long (~17-18 kDa) should be produced, followed by the maturation of the 11 kDa protease (pathway B).

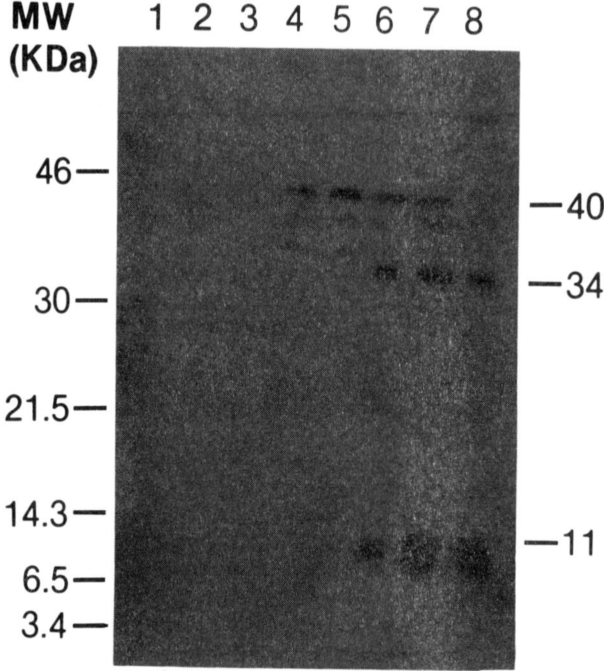

Figure 3.
Biosynthesis and autoprocessing of HIV-1 protease in E. coli L507 cells transformed by plasmid pHP10 (L5-10 strain). Western blot analysis of HIV-1 protease produced in L5-10 cells. Bacteria were grown to mid log phase at 32°C. Gene expression was induced by shifting cultures to 40°C. One millilitre sampler were taken at times indicated, chilled and pelleted. Cells equivalent to 0.15 OD_{600} taken at each time point were analyzed by SDS-PAGE (15%) and immunoblotting, using rabbit antiserum raised against a synthetic peptide corresponding to HIV-1 protease residues 17-36. The antigen antibody complexes were detected by ^{125}I protein A (Amersham).

Our results, as depicted in Figure 3, indicated that in L5-10 cells, autocatalysis of HIV-1 protease followed pathway A, that is to say that in this expression system, HIV-1 protease processed itself first at its N terminus. This conclusion agrees with the one reached by Stickler *et al.*. (1989), but contradicts with the one drawn by Hansen *et al.*. (1988), using similar prokaryotic expression systems. The expression vector used by Hansen and coworkers to teach their conclusion also contained a *Bgl* II-*Eco* RV fragment of HIV-1 DNA. This fragment, missing the first 4 codons of *pol* ORF, was fused upstream to the translation initiation signal and the first 98 codons of MS2 polymerase gene. Their insert encodes a primary translation product 45 kDa in size comprising 98 aa of MS2 *pol* protein, 52 aa of L region, 99 aa of mature protease and 144 aa of N terminal reverse transcriptase. The difference lies in the positions of the heterologous genes fused to the HIV-1 protease coding sequence. It is

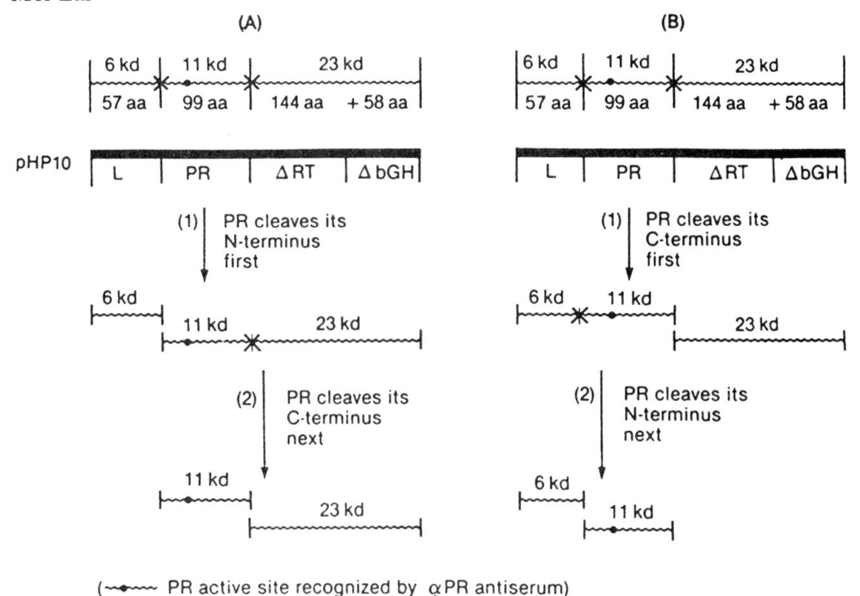

Figure 4.
Two possible pathways for autoprocessing of HIV-1 protease in the *E. coli* cells transformed by pHP10.

likely that MS2 polymerase placed upstream and bGH placed downstream may affect the conformation of the polyproteins, there by influencing the autoprocessing pattern of the protease. Loeb and coworkers (Loeb *et al.*, 1989) recently reported the HIV-1 protease cannot process itself at either the amino terminus or the carboxyl terminus if the last residue of HIV-1 protease, Phe 99 , is substituted by residues that are not tolerated. Hostomsky *et al.*, (1989) also reported that a mutant with Phe 99 deleted is unable to cleave at the N terminus of the enzyme. Based on these observations, both groups suggested that autocatalysis of HIV-1 protease must take place first at the carboxyl terminus. However, their results may also be interpreted to mean that Phe 99 is critical for the molecular conformation essential for the enzyme to cleave at the N terminal site. In fact, the hydrogen bonding between proline at the N terminus of one monomer and Phe 99 of the second monomer that constitutes a portion of a chain at the inner part of the intra-dimer interface, as revealed by the crystal structure of HIV-1 protease (Wlodawer *et al.*, 1989), may lend support to this notion.

Processing of the *gag-pol* Polyprotein

To further examine the biosynthesis and processing of the *gag* and *pol* proteins of HIV-1, two additional expression vectors were constructed. Plasmid pPG (Figure 5) was derived from pHP10 by joining the *gag* sequence form nt789 to nt2095 (*Bgl* II site) upstream of the insert in pHP10 at the *Bgl* II site. The insert in pPG included nt789-2978 of clone HXB2 and the last 58 codons of bGH. The *pol* ORF in this plasmid was in the minus one position in relation to the *gag* ORF, exactly as existed in the viral genome. The insert in pPG encoded the entire *gag-poly*protein (500 aa, pr55^{gag}, nt789-2288) and 358 aa derived from the insert in pHP10 in translated in minus one frame. Plasmid pCPG (Figure 5) was constructed from pPG by digesting pPG with *Bgl* II (nt2095, near the frameshift site required for the synthesis of pr160$^{gag-pol}$ fusion protein), followed by Klenow fill in reaction. This manipulation resulted in adding 4 nucleotides to pPG at the *Bgl* II site.

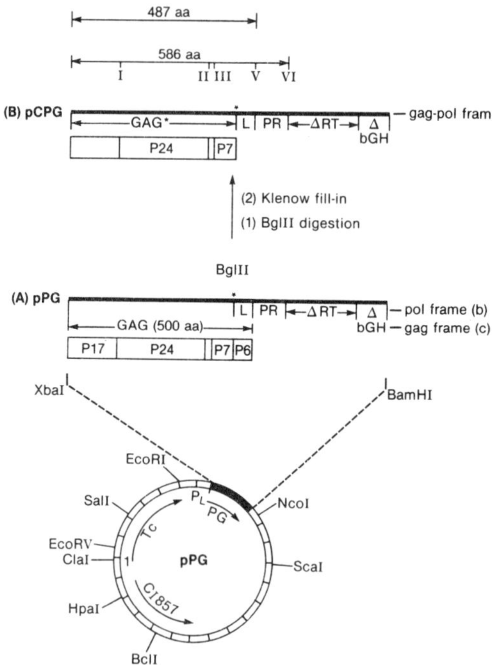

Figure 5.
Plasmid maps of pPg and pCPG. Plasmid pPG was derived from pHP10 (Fig. 2) by linking the *gag* gene upstream to the insert in pHP10. Plasmid pCPG was constructed from pPG by inserting 4 base pairs at the *Bgl* II site in the *gag-pol* overlapping region. See text for details.

To study the expression of these sequences in *E. coli*, L507 cells transformed by pPG and pCPG were grown and induced as described for pHP10 transformed cells. Samples were taken at 0, 2.5, 10, 20, 40, 60 and 90 minutes after temperature shift up, and analyzed by SDS PAGE and Western analysis, using a mixture of monoclonal antibodies against *gag* proteins p17(MA) and p24(CA). The results are depicted in Figure 6. In pPG transformed cells, the expected primary translational product of the *gag* gene, pr55^{gag} , was detected as early as 2.5 minutes postinduction (Figure. 6, lane 3). Its synthesis continued to increase up to 60 minutes and then levelled off (Figure 6, lanes 5, 7, 9, 11 and 13). A second major species of *gag* related protein of ~48 kDa in size appeared 20 minutes (Figure 6, lane 7) after induction. Its production continued unabated even at the end of the induction period (Figure 6,lanes 9, 11 and 13). This 48 kDa protein could be generated either by bacterial degradation of pr55*gag* , or via internal initiation at the second ATG codon at residue 142 of *gag* polyprotein (in p24). The latter seemed unlikely because both p17 and p24 antibodies reacted with p48 (data not shown) and its size was too large for a protein initiated at Met142 codon.

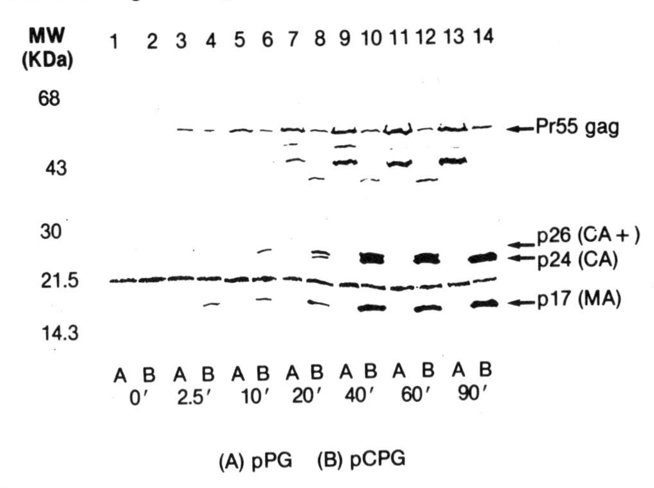

Figure 6.
Biosynthesis and processing of *gag* proteins in *E. coli* transformed by plasmid pPG (L5 PG strain) and by pCPG (L5 CPG strain). Immunoblotting analysis of synthesis and processing of *gag* proteins in L5 PG cells. Cell growth, induction of gene expression, and analysis of samples was performed as described in legend to Figure 3. The *gag* related proteins were detected by monoclonal antibodies against p17 and p24 (Dupont) and immunoperoxidase ABC technique (Vectastain ABC Kit, Vector Laboratories).

For p17 and p24 to be cleaved from pr55*gag*, a functional protease must be made from the *gag-pol* fusion protein generated by a ribosomal frameshift event. In pPG transformed cells, the accumulation of pr55*gag* and the absence of p17 and p24 indicated that an active HIV-1 protease was not produced. This was not due to lack of protease coding mRNA (data not shown). Rather, it probably resulted from the inability of *E. coli* protein synthesizing machinery to recognize the frameshift signal of HIV-1 RNA.

As mentioned earlier, the insert in pCPG was identical to that in pPG with the exception of 4 bp filled in near the frameshift site. This addition fused the *gag* ORF and the *pol* ORF into the same reading frame. The amino acid sequence of the experimentally fused *gag-pol* polyprotein was nearly identical to that of *gag-pol* protein achieved by frameshifting, except that the first 4 codons (Phe-Phe-Arg-Glu) of the *pol* protein were replaced by *gag* residues 433-437 (Phe-Leu-Gly-Lys-Ile). This change should not affect the autoprocessing of HIV-1 protease, since it has been reported that only 4 residues surrounding the N terminal cleavage site (site V) are required for correct autocatalysis (Loeb *et al.*, 1989).

The temporal sequence of processing of *gag* and *pol* proteins at sites I, II, III, V and VI was investigated in E. coli transformed by pCPG. The induction experiment was performed in parallel with that of pPG transformed cells. The results are depicted in Figure 6 (lanes 2, 4, 6, 8, 10, 12 and 14). The fused *gag* and truncated *pol* genes in pCPG had a coding capacity for 791 aa (~87 kDa, termed $pr87^{gag-\Delta pol}$). However, $pr87^{gag-\Delta pol}$ was never detected throughout the experimental course. Instead, the largest *gag* related protein, detected as early as 2.5 minutes postinduction (Fig. 6, lane 4), was a 55 kDa protein that comigrated with $pr55^{gag}$ encoded by the entire *gag* ORF in pPG transformed cells. This 55 kDa protein most likely represented the protein generated by protease autoprocessing at site V (N terminus of protease) located 13 aa upstream from the carboxyl terminus of $pr55^{gag}$. The difference of 13 aa (500 aa vs. 487 aa) would be nearly indistinguishable in the 15% gel employed in this study. If autoprocessing occurred first at the carboxyl end of the protease, a 586 aa *gag* related protein (~65 kDa) should be generated. Such a protein, however, was never detected in pCPG transformed cells. These results again indicated that HIV-1 protease apparently processed itself at its amino terminus first, followed by cleavage at the carboxyl terminus to generate the mature protease. Although p17 appeared as early as 2.5 minutes after induction (Figure 6, lane 4), cleavage at site I to generate p17 probably required a mature protease to act *in trans*, and therefore mist be preceded by cleavages at sites V and VI. At 10 minutes after induction, a 26 kDa protein (reactive with p24) was detected, corresponding to a protein generated by cleavage at sit III. The 26 kDa protein further matured to give rise to p24 (Fig. 6, lanes 6, 8, 10, 12, 14) after cleavage at site II. Mervis and coworkers had made similar observations in A3.01 cells infected with HIV-1 (Mervis *et al.*, 1988). We concluded that in pCPG transformed E. coli cells, the processing of $pr87^{gag\ pol}$ fusion occurred first at site V, followed in order by cleavage at site VI, site I, site III and site II. It is unclear whether processing of *gag-pol* proteins must follow a specific pathway. Nonetheless, this temporal sequence appeared to be the predominant one in E. coli transformed by plasmid pCPG.

ACKNOWLEDGEMENTS

We thank F. Ruscetti (NCI) for clone HXB2, H. Hsiung for vector pL110c, S. R. Jaskunas for L507 cells, and B. Fogleman for manuscript preparation.

REFERENCES

Chassagne, J., Verrelle, P., Dionet, C., Clavel, F., Barre Sinoussi, F., Chermann, J.C., Montagnier, L., Gluckman, J.C. and Klatzmann, D. *J. Immunol.*, **136**, 1442-1445, (1986).

Darke, P.L., Nutt, R.F., Brady, S.F., Garsky, V.M., Ciccarone, T.M., Leu, C. T., Lumma, P.K., Freidinger, R.M., Veber, D.F. and Sigal, I.S. *Biochem. Biophys. Res. Comm.*, **156**, 297-303, (1988).

Farmerie, W.G., Loeb, D.H., Casavant, N.C., Hutchinson III, C.A., Edgell, M.H. and Swanstrom, R. *Science*, **236**, 305-308, (1987).

Gendelman, H.E., Theodore, T.S., Willey, R., McCoy, J., Adachi, A., Mervis, R.J., Venkatesan, S. and Martin, M.A. *Virology*, **160**, 323-329, (1987).

Graves, M.C., Lim, J.J., Heimer, E.P. and Kramer, R.A. *Proc. Acad. Sci. USA*, **85**, 2449-2453, (1988).

Hansen, J., Billich. S., Schulze, T., Sukrow, S. and Moelling, K. *EMBO J.*, **7**, 1785-1791, (1988).

Henderson, L.E., Copeland, T.D., Sowder, R.C., Schultz, A.M. and Oroszlan, S. *UCLA Symp. Mol. Cell. Biol.*, **71**, 135-147, (1988).

Hostomsky, A., Appelt, K. and Ogden, R.C. *Biochem. Biophys. Res. Comm.*, **161**, 1056-1063, (1989).

Jacks, T., Power, M.D., Masiarz, F.R., Luciw, P.A., Barr, P.J. and Varmus, H.E. *Nature*, **331**, 280-283, (1988).

Kalyanaraman, V.S., Cabradilla, C.D., Getchell, J.P., Narayanan, R., Braff, E.H., Chermann, J.C., Barre Sinoussi, F., Montagnier, L., Spira, T.J., Kaplan, J., Fishbein, D., Jaffe, H.W., Curran, J.W. and Francis, D.P. *Science*, **225**, 321-323, (1984).

Kohl, N.E., Emini, E.A., Schleif, W.A., Davis, L.J., Heimbach, J.C., Dixon, R.A.F., Scolnick, E.M. and Sigal, I.S. *Proc. Natl. Acad. Sci. USA*, **85**, 4686-4690, (1988).

Kramer, R.A., Schaber, M.D., Ganguly, K., Wong Staal, F. and Reddy, E.P. *Science*, **231**, 1580-1584, (1986).

Lightfoote, M.M., Coligan, J.E., Folks, T.M., Fauci, A.S., Martin, M.A. and Venkatesan, S. *J. Virol.*, **60**, 771-775, (1986).

Lillehoj, E.P., Rick Salazar, F.H., Mervis, R.J., Raum, M.G., Chan, H.W., Ahmad, N. and Venkatesan, S. *J. Virol.*, **62**, 3053-3058, (1988).

Loeb, D.D., Hutchinson III, C.A., Edgell, M.H., Farmerie, W.G. and Swanstrom, R. *J. Virol.*, **63**, 11 121, (1989).

Mervis, R.J., Ahamd, N., Lillehoj, E.P., Raum, M.G., Salazar, F.H.R., Chan, H.W. and Venkatesan, S. *J. Virol.*, **62**, 3993-4002, (1988).

Pearl, L.H. and Taylor, W.R. *Nature*, **329**, 351-354, (1987).

Ratner, L., Haseltine, W., Patarca, R., Livak, K.J., Starcich, B., Josephs, S.F., Doran, E.R., Rafalski, J.A., Whitehorn, E.A., Baumeister, K., Ivanoff, L., Petteway, S.R., Pearson, M.L., Lautenberger, J.A., Papas, T.S., Ghrayeb, J., Chang, N.T., Gallo, R.C. and Wong Staal, F. *Nature*, **313**, 277-284, (1985).

Sarngadharan, M.G., Bruch, L., Popovic, M. and Gallo, R.C. *Proc. Natl. Acad. Sci. USA*, **82**, 3481-3484, (1985).

Stickler, J.E., Gorniak, J., Dayton, B., Meek, T., Moore, M., Magaard, V., Malinowski, J. and Debouck, C. *Proteins*, **6**, 139-154, (1989).

Wlodawer, A., Miller, M., Jaskolski, M., Sathyanarayana, B.K., Baldwin, E., Weber, I.T., Selk, L.M., Clawson, L., Schneider, J. and Kent, S.P.H. *Science*, **245**, 616-621, (1989).

8
Dimerisation of the HIV-1 Protease: Preliminary Analysis Using Gel Permeation Chromatography

O. M. P. Singh, E. M. J. Roud Mayne and M. P. Weir

The role of HIV-1 virus as the causative agent of AIDS and related disorders has been clearly demonstrated (Gallo *et al.*, 1984). Virus encoded proteinase is involved in the processing of *gag* and *gag-pol* polyproteins (Kohl *et al.*, 1988) which allows the virus to mature and subsequently cause infection. Pepstatin A and its analogues along with various inhibitors of renin are able to inhibit the proteinase activity, thus demonstrating that the HIV-1 proteinase belongs to the aspartyl protease family (Richards *et al.*, 1989).

Based on sequence homology and secondary structure predictions the active enzyme was expected to form homodimeric structures (Pearl and Taylor, 1987). The recently determined 3-D structure confirms that the enzyme is indeed dimeric (Wlodawer *et al.*, 1989; Lapatto *et al.*, 1989) and similar to the Rous Sarcoma virus proteinase (Miller *et al.*, 1989). Furthermore, the topological arrangement of the two subunits demonstrates an evolutionary kinship to other aspartyl proteases. Solution studies also indicate that the active form of the enzyme is dimeric; monomeric enzyme was not observed in these studies (Nutt *et al.*, 1988; Meek *et al.*, 1989).

Strategies for intervention in the virus life cycle through inhibition of the proteinase are promising. Apart from the active site titrants as inhibitors another approach is to prevent the protein dimerisation.

METHODS

We have expressed the BH10 isolate encoded proteinase (*gag-pol* residues 1-168) in *E. coli* as a cleavable, truncated chloramphenicol acetyltransferase fusion using the IPTG inducible tac promoter (Singh *et al.*, in preparation). Soluble enzyme from these cells has been purified in high yield by acetone precipitation, ion exchange, hydrophobic interaction and gel permeation chromatography. Details of assay conditions and gel permeation chromatography are given in the figure legends. The cleavage of the substrate peptide, IGCTLNFPISPIETV (form the C-terminus of proteinase and N-terminus of reverse transcriptase) was determined from the reverse phase separation on PepRp HR 5/5 column (Pharmacia, UK). The N and C terminus fragments and the substrate were resolved by using a TFA/acetonitrile gradient from 15-40% at a flow rate of

1ml/min. Both fragments have been verified by composition analysis. The system was calibrated using known amounts of the C-terminus fragment. Gel permeation chromatography was done using Superdex G75 column (1.6 x 70cm) which was kindly loaned to us by Pharmacia UK. Various conditions used for the different separations are described in figure legends.

RESULTS AND DISCUSSION

During our studies on the optimisation of the enzyme assay, it appeared that the enzyme catalysis was very sensitive to the buffer ionic strength. Detailed analysis of this indicated that the enzyme reaction rate can increase up to 4 fold by increasing NaCl from 0-2.0M. (Figure 1). This finding is similar to that reported for the avian myeloblastosis virus proteinase as well as pepsin (Nutt *et al.*, 1988). Whilst it is difficult to

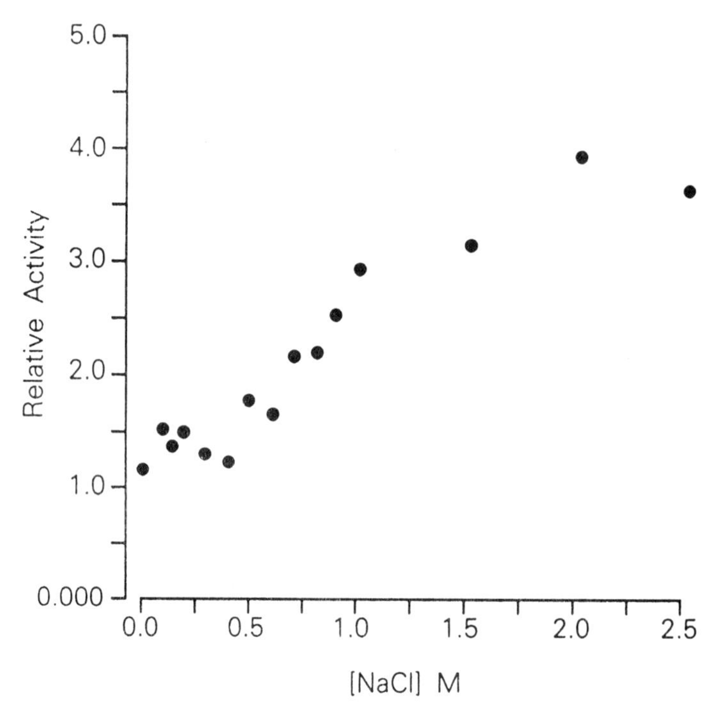

Figure 1.
Effect of increasing NaCl concentration on the proteinase activity. Partially purified enzyme was assayed in the presence of 0-2.0M NaCl, 150μM peptide substrate, 2mM EDTA, 50mM Morpholinoethane sulphonic acid buffer at pH 6.0 in a total volume of 200ul. After incubation at 37°C for 1min, the samples were boiled for 5min, centrifuged and 100μl loaded on the reverse phase column. Activity increase relative to that found at 0.0M NaCl (571nmoles/min/ml) is reported.

interpret the effect of NaCl on pepsin one likely cause for the increase in the AMV and HIV-1 proteinase activity could well be due to altered dimerisation. Alternatively it could increase the substrate affinity to the enzyme, thereby decreasing the Km and increasing the V_{max}.

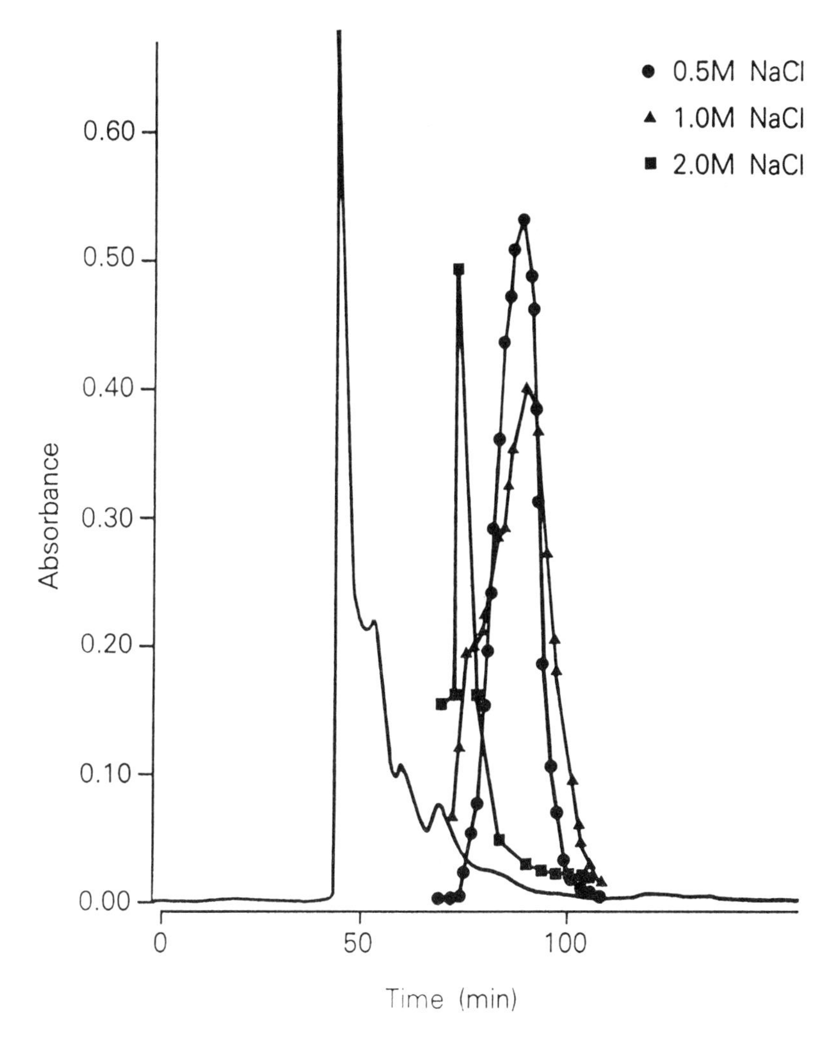

Figure 2.
Gel permeation chromatography in different NaCl concentrations and its consequences on the proteinase elution. Approx. 25-50µg of active proteinase in 0.2ml volume was applied to the Superdex G75 column which was equilibrated in 0.5-2.0M NaCl, 10% glycerol, 2mM EDTA and 50mM morpholino-ethane sulphonic acid buffer at pH 6.0 and at 4°C. 50µl of the eluate from each fraction was assayed as described.

We set out to observe dimerisation directly by gel permeation chromatography. Small amounts of partially purified enzyme (post ion-exchange and hydrophobic interaction chromatography; 25–50μg of proteinase) was separated on Superdex G75. When the column is developed in buffer containing 0.5M NaCl then the activity appears to elute in fraction corresponding to molecular weight of 9K thus suggesting that in this environment majority of the protein population is in the monomeric form. Increase in NaCl to 1.0M results in a separation of the main activity peak into two partially resolved components with the main component corresponding to the monomer whereas the second component appeared to correspond to a protein molecular weight of 20K, thus suggesting that this new population is representative of the dimeric form of the enzyme. Further chromatography in the presence of 2.0M NaCl was attempted. This time there was only one main peak of activity corresponding to molecular weight of 20K (Figure 2). The calibration curve for the Superdex G75 column was linear and used to estimate the molecular size of the different populations of the proteinase (Figure 3).

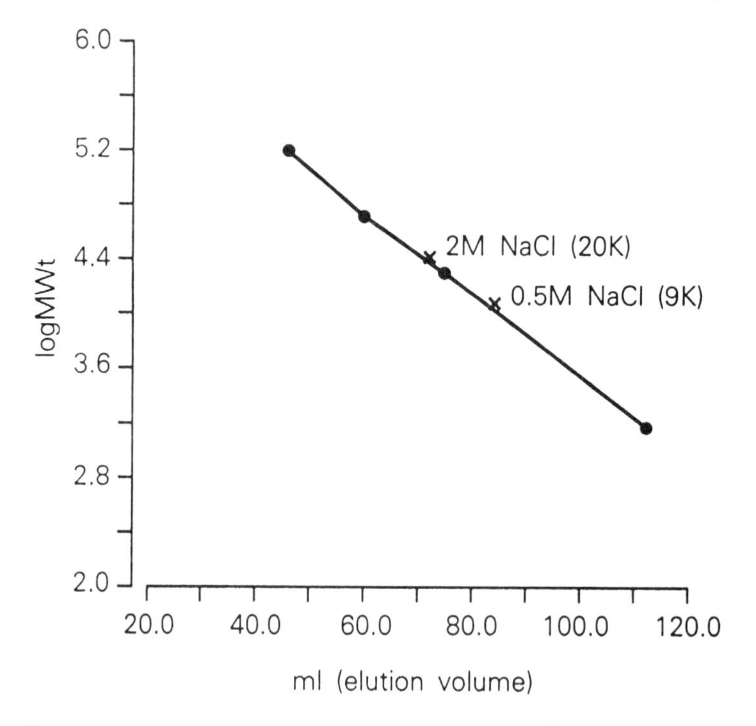

Figure 3.
Estimation of the proteinase molecular weight. Gel filtration standard (Bio-Rad, UK) was dissolved in 1ml buffer and 100ul applied to the column. The separation conditions were as described. Bovine thyroglobulin (670kD) and gamma globulin (158kD) eluted together, chicken ovalbumin (44kD), horse myoglobulin (17kD) and vitamin B12 (1.3kD) were clearly resolved.

Adsorption to the chromatographic matrix is observed with TSK2000SW gel (Strickler *et al.*, 1989) and is seen with Superdex at 0M NaCl where the apparent molecular weight is only 1-2K ; however, adsorption does not explain the presence of both species at 1M NaCl.

Thus it appears as if, like the AMV enzyme activity, HIV-1 proteinase is also sensitive to its ionic environment. The fact that NaCl concentration as high as 2.0M appears to assist dimerisation would suggest that the likely interactions between the dimer are quite weak. The enzyme is active in conditions where the gel permeation chromatography indicates the protein to be in the monomeric form, the substrate must then be able to induce and stabilise the dimer form. Whilst there is no evidence at present which demonstrates that the protease within the mature virus is a stable dimer then the possibility of a stable monomer cannot be discarded. Most of the members of the aspartyl protease family are much larger, the fact that the retroviruses carry only one domain must confer some advantage. Apart from reduced load on the retrovirus, the monomer may well allow specific cleavages between the 'folded' *gag* and *gag-pol* components by the monomer recognising the linking short stretches first followed by dimerisation and cleavage. Alternatively,, the microenvironment within the virion may be such that the enzyme forms tightly associated dimers. The pH within the virion is not known, concievably there is a lowering of pH to allow for optimum catalysis rate to allow efficient maturation.

From the evidence of the enzyme's behaviour in NaCl, the separation of monomer and dimer on gel permeation chromatography and the cross linking data (Katoh *et al.*, 1989) from RSV which suggests monomer and dimer mixtures in solution, along with the demonstration that CNBr fragments of the RSV enzyme affect dimerisation (Katoh *et al.*, 1989) we feel that specific dimerisation inhibitors could be found.

REFERENCES

Gallo, R.C., Salahuddin, S.Z., Popovic, M., Shearer, G.M., Kaplan, M., Haynes, B.F., Palker, T.J., Redfield, R., Oleske, J., Safai, B., White, G., Foster, P. and Markham, P. *Science*, **224**, 500-502, (1984).

Katoh, I., Ikawa, Y. and Yoshinaka, Y. (1989) *J. Virol.*, **63**, 2226-2232, (1989).

Kohl, N.E., Emini, E.A., Schleif, W.A., Davis, L.J., Heimbach, J.C., Dixon, R.A.F., Scolnick, E.M., Sigal, I.S. *Proc. Natl. Acad. Sci. USA*, **85**, 4686-4690, (1988).

Kotler, M., Katz, R., Danho, W., Leis, J. and Skalka, A. *Proc. Natl. Acad. Sci. USA*, **85**, 4185-4189, (1988).

Lapatto, R., Blundell, T., Hemmings, A., Overington, J., Wilderspin, A., Wood, S., Merson, J.R., Whittle, P.J., Danley, D.E., Geoghegen, K.F., Hawrylik, S.J., Lee, S.E., Scheld, K.G. and Hobart, P.M. *Nature*, **342**, 299-302, (1989).

Meek, T.D., Dayton, B.D., Metcalf, B.W., Dreyer, G.B., Strickler, J.E., Gorniak, J.G., Rosenberg, M., Moore, M.L., Magaard, V.W. and Debouck, C. *Proc. Natl. Acad. Sci. USA*, **86**, 1841-1845, (1989).

Miller, M., Jaskolski, M., Rao, J.K.M., Leis, J. and Wlodawer, A. *Nature*, **337**, 576-579, (1989).

Nutt, R.F., Brady, S.F., Darke,P.L., Ciccarone, T.M., Colton, C., Nutt, E.M., Rodkey,J.A., Bennett, C.A., Waxman, L.H., Sigal, I.S., Anderson, P.S. and Veber, D.F. *Proc. Natl. Acad. Sci. USA*, **85**, 7129–7133, (1988).

Pearl, L.H. and Taylor, W.R. *Nature*, **329**, 351–354, (1987).

Richards, A.D., Roberts, R., Dunn, B.D., Graves, M.C. and Kay, J. *FEBS Lett.*, **247**, 113–117, (1989).

Singh, O.M.P., Montgomery, D.S., Roud Mayne, E.M.J., Weir, M.P.,Dykes, C.W. and Hobden, A.N. (In preparation)

Strickler, J.E., Gorniak, J.G., Dayton, B.D., Meek, T.D., Moore, M.L., Magaard, V.W., Malinowski, J. and Debouck, C. (1989) *Proteins*, **6**, 139–154, (1989).

Wlodawer, A., Miller, M., Jaskolski, M., Sathyanarayana, B.K., Baldwin, E., Weber, I.T., Selk, L.M., Clawson, L., Schneider, J. and Kent, S.B.H. *Science*, **245**, 616–621, (1989).

9
Three-dimensional Structure and Evolution of HIV-1 Protease

Andrew Wilderspin, Duncan Gaskin, Risto Lapatto, Tom Blundell, Andrew Hemmings, John Overington, Jim Pitts, Stephen Wood, Zhang-Yang Zhu, Laurence H. Pearl, Dennis E. Danley, Kieran F. Geoghegan, Stephen Hawrylik, S. Edward Lee, Kathryn Shield, Peter M. Hobart, James Merson and Peter Whittle

HIV-1 proteinase processes its virally encoded polyproteins into mature structural proteins and enzymes that are essential for viral propagation. As a consequence the proteinase is an attractive target for prospective antiviral agents for the treatment of AIDS, and knowledge of its tertiary structure an important step in drug design. Following the observation (Toh *et al.* 1985) that retroviral proteinases shared a highly conserved sequence Asp-Thr/Ser-Gly with the pepsins, it has been hypothesised (Pearl and Taylor, 1987; Blundell *et al.* 1988) on the basis of sequence analysis and modelling studies that these enzymes exist as dimers closely similar in three-dimensional structure to the ancestral dimeric proteinase suggested for the aspartic proteinases (Tang *et al.* 1978). This has now been confirmed, first by X-ray analysis of a synthetic HIV-1 proteinase in the laboratory of Wlodawer (Weber *et al.* 1989) and then for a recombinant enzyme in our own laboratories (Lapatto *et al.* 1989). These X-ray structure analyses indicated that the overall fold of the HIV-1 proteinase closely resembled that of the RSV-proteinase (Miller *et al.*, 1989). The Asp-Thr-Gly sequences adopt a conformation closely similar to that of the pepsin-like aspartic proteinases but organised symmetrically in the dimer about the crystallographic 2-fold axis. However, the N- and C-termini together form an intermolecular four-stranded sheet, which is central to the stability of the dimer, in contrast to the inter-subunit sheet of the pepsins, which has six antiparallel strands arranged around the pseudo dyad.

In this paper we review our recombinant DNA studies and describe the structure of the recombinant enzyme defined by X-ray analysis. We compare the active sites and specificity pockets of the retroviral aspartic proteinases. We show by analysis of the sequences and by modelling that the retroviral proteinase structure provides a satisfactory explanation for the invariance or conservative variation of amino acid residues amongst retroviral proteinases.

Recombinant studies at Birkbeck.

At Birkbeck we pursued two different strategies to produce recombinant enzyme in sufficient quantitites for structural studies. Initially we constructed a synthetic gene based on the amino acid sequence derived from the BH10 clone (Ratner *et al*, 1985). The synthetic gene was designed to encode a methionine fused to the 99 residues of the mature proteinase. The codon usage was optimised for expression in *E. coli* and the gene carried its own Shine-Dalgarno sequence situated eight bases upstream of the methionine start codon.

Sixteen oligonucleotides encoding both strands of the gene were synthesized on a 0.2 μmole scale with an Applied Biosystems 381A DNA Synthesizer using β-cyanoethyl protected phosphoramidite chemistry. After purification by reverse-phase chromatography, the oligonucleotides were detritylated, the 5' ends phosphorylated and then, in a single reaction, annealed together. The ligated gene, inserted into the M13mp18 vector, was found to have the correct sequence. We explored expression in *E. coli* using the plasmid pKK223-2, which places the gene under the control of the *tac* promoter. Although a 10 kDa band was not seen on polyacrylamide gel electrophoresis, cleavage of an octapeptide, SFNFPQIT (which corresponds to the N- terminal cleavage sequence of the proteinase) could be observed in an HPLC assay system.

A sub-clone of the wild-type sequence of BH10 has also been used in experiments to fuse a portion of the pol gene to a number of genes that promote secretion or inclusion body formation in *E. coli*. Such schemes may avoid the apparent problems of proteinase toxicity to the cell and enable production of protein from a small fermentation. A gene encoding a C-terminal fusion of the proteinase to the 23 kDa *E. coli* haemolysin signal sequence was constructed, but co-expression with the translocation function resulted in secretion of the signal sequence only, suggesting autocatalytic cleavage by the proteinase prior to secretion. Similarly we have found that inclusion bodies produced from constructs expressing half of the chymosin gene ligated to the HIV-1 *pol* gene contain mainly truncated chymosin indicating that autocatalytic cleavage is again occurring.

Recombinant studies at Pfizer.

At Pfizer we were more successful in providing HIV-1 proteinase for crystallisation. A recombinant fragment of the viral *pol* gene was expressed in *E. coli*. The bacterial expression plasmid pHIVexpol5 was constructed by subcloning the 5.0 kb *Bgl* II fragment from the pHXBc2 clone of the viral genome (Ratner *et al.*, 1985) into a *Bgl* II - cleaved derivative of pPFZ-R2 (Franke, 1989). This yielded a gene controlled by the *trp* promoter and encoding a fusion polypeptide comprised of an initiator methionine, a seven-residue junction sequence -Ala-Glu-Ile-Thr-Arg-Ile-Glu-, and a 115 kDa fragment of the *pol* gene product which commenced with Asp 5 and encompassed the proteinase, reverse transcriptase and integrase domains.

E. coli MM294 cells harbouring pHIVexpol5 were shown by Western blotting and immunodetection to produce appropriately processed proteinase (11 kDa), reverse transcriptase (66 and 51 kDa forms), and integrase (34 kDa). Proteinase activity, already implied by the appropriate intracellular processing of the *pol*-derived polyprotein, was further demonstrated as the ability of cell lysates to effect specific cleavage of peptide substrates. The lysates were also active in assays conducted using recombinant-derived *gag* protein as substrate. HIV-1 proteinase was purified from the soluble fraction of cell lysates in four steps of chromatography (similar to that used in Darke *et al.*, (1989)) to give a single band on Coomassie blue-stained SDS-polyacrylamide gels. Automated Edman degradation gave an N-terminal sequence of Pro-Gln-Ile-Thr-Leu-, consistent with the expected result of self-processing of the pol gene product; in addition, sequencing revealed the presence of no protein sequence other than that of the proteinase. 0.25 mg of purified proteinase was recovered per litre of bacterial culture. The specific activity of the purified enzyme was measured using a 17-residue synthetic peptide substrate (RRVTNSATIMMQRGNFR; sequence suggested by Dr J Schneider), and was estimated at 2.4 μmol substrate cleaved/min/mg of proteinase (pH 5.5, 23°C, 0.2 mM substrate). Details of the expression studies, cell growth and protein purification will be presented elsewhere (Hobart *et al.* 1990).

Crystallisation and X-ray analysis.

Crystals shown in Figure 1 were prepared using the published protocol (McKleever *et al.*, 1989) and grew at 4°C in five days. They were isomorphous with those reported earlier (Navia *et al.*, 1989) of space group $P4_12_12$. FAST area detector data collection over a period of two days allowed 10,000 reflections to be recorded. A merging R-value was 0.11 leading to over 2200 reflections with $I > 2.0\sigma(I)$ with a maximum resolution of 2.7Å resolution. We have also collected a partial data set using synchrotron X-radiation to a resolution of 2.2Å.

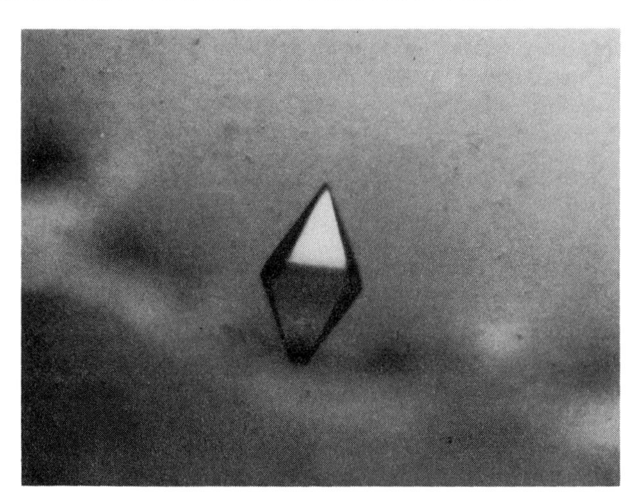

Figure 1.
Crystals of HIV1-proteinase prepared using the published protocol (McKleever et al., 1989) and grew at 4°C in five days. They were isomorphous with those reported earlier (Navia *et al.*, 1989) of space group $P4_12_12$.

We had earlier constructed models using Composer (Blundell *et al.* 1988) initially from the pepsin-like aspartic proteinase domains and later from the RSV structure. From these we constructed a model comprising residues 12 to 46 and 55 to 84, that is only those strands that were common to HIV and RSV structures reported earlier (Miller *et al.* 1989; Navia *et al.* 1989). Thus the initial model used did not include the N- or C-terminal strands, the helix or the flap, the position of which were in contention. The position of this model was defined by molecular replacement, checked with respect to the coordinates of the HIV model (Navia *et al.* 1989) and refined using rigid body and restrained least squares. A series of difference Fourier analyses, modelling and refinement steps gave first the region 47 to 54 (the flap), then the N- and C-termini, and finally the turn after the N-terminal strand (residues 5 to 11) and the α-helix (87 to 93). In each case the polypeptide could be followed in the difference electron density (calculated on the basis of a model in which they had not previously been included) without a break and with clear positions for the sidechains. Thus we achieved an unequivocal tracing of all 99 amino acids of the polypeptide chain.

Refinement of the initial model using rigid group refinement was achieved using RESTRAIN (Haneef *et al.*, 1985). In the final cycles individual thermal paramenters but tight geometric restraints were used. Electron densities were calculated from difference Fouriers with coefficients $2F_{obs}-F_{calc}$ and displayed simultaneously with positive and negative contours for electron densities for difference Fouriers calculated with coefficients $F_{obs}-F_{calc}$ using FRODO (Jones, 1978) on an Evans and Sutherland PS390. The model was extended in steps to include the N- and C- termini and the helix in steps and at each stage refined using restrained least squares refinement using RESTRAIN.The final R-value is 0.189.

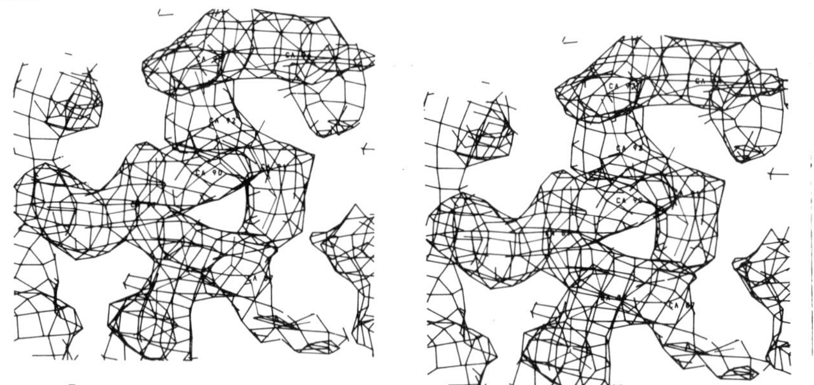

Figure 2.
Stereo views of a difference electron density maps calculated with coefficients $2F_{obs}-F_{calc}$ for residues 87 to 90 which form part of the helix, *h'*, preceding the terminal strand, *t*. These atoms were omitted from the calculation of structure factors. The strands and helices are defined in the legend to Figure 5.

Figure 2 shows the difference electron density for a region where our interpretation differs from that of Navia *et al* (1989). Figure 3 shows electron density based on a Fourier with coefficients $2F_{obs}-F_{calc}$ and based on the final refined coordinates.

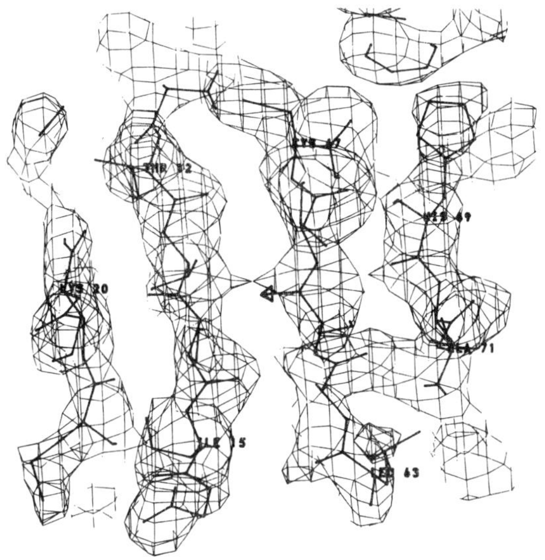

Figure 3.
shows electron density based on a Fourier with coefficients $2F_{obs}-F_{calc}$ and based on the final refined coordinates.

The three-dimensional structure.

Figure 4 shows stereo views of the three-dimensional structure of the enzyme which exists as a dimer with a well defined and extensive active site cleft. Figure 5a shows a two-dimensional representation of the secondary structure and hydrogen bonding in the dimeric structure. This is compared with that of a pepsin-like aspartic proteinase shown in Figure 5b. The arrangements are strikingly similar and thus we adopt the nomenclature (Blundell *et al.* 1985) for the strands for the aspartic proteinase, endothiapepsin. Each subunit comprises two similar motifs related by an approximate two-fold axis; each motif contains anti-parallel strands: *a*, *b*, *c* and *d* for the first and *a'*, *b'*, *c'* and *d'* for the second. These are organised together in a distorted sheet with strands *c* and *d'* and strands *c'* and *d* forming two pairs of parallel strands (see figure 5). Strands *b* and *c* and strands *b'* and *c'* form antiparallel beta-hairpins that are folded on top of the first sheet and hydrogen-bonded together around the same two-fold axis that relates the motifs in the main sheet. The strands *d* and *d'* are followed by an irregular region and an α-helix respectively that also lie on this side of the main mixed β-sheet. Strand *a* of the first motif is displaced from the main mixed sheet and forms an anti-parallel four-stranded β-sheet with the terminal strand, *t*, and the two equivalent strands from the other subunit of the dimer. Figure 4b shows that these strands lie on the other side of the dimer from the well defined cleft. These strands occupy the same volume but have different orientations to the six stranded β-pleated sheet that connects the two domains of the aspartic proteinases (strands *a*, *r* and *q* of the N- and C-terminal domains in Figure 5b). This corresponds closely to the arrangement of the completely symmetrical dimer that was proposed as the ancestor of the aspartic proteinases (Tang *et al.* 1978).

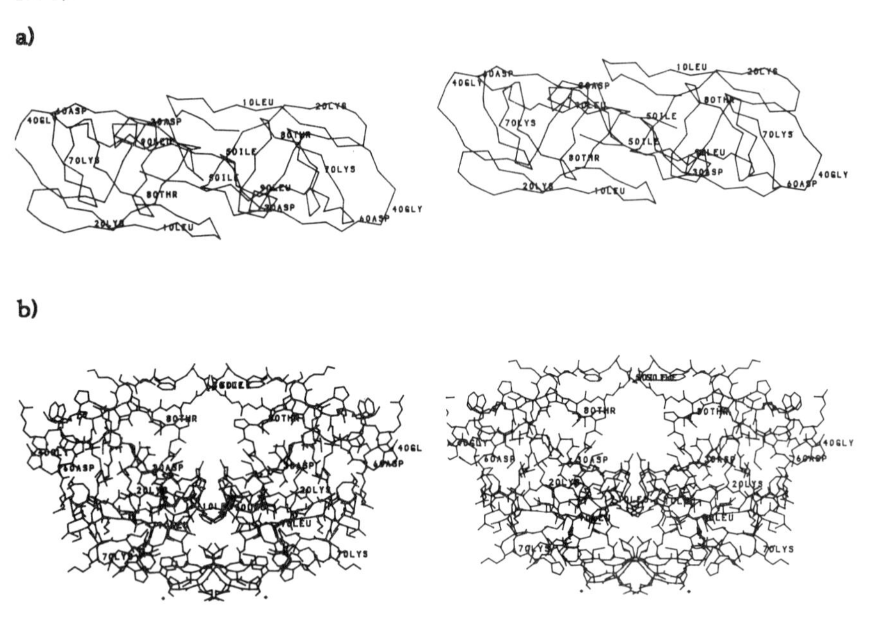

Figure 4.
Stereo views of (a) Cα-atoms viewed from the side of the dimer containing the four-stranded β-sheet along the two-fold axis relating subunits in the dimer, and (b) all atoms viewed perpendicular to the two-fold and along the active site cleft.

There is an extensive hydrophobic core that extends from one subunit through the dimeric interface to the other subunit. The residues involved in the core are indicated as squares in Figure 5. Their identities can be derived from Figure 6 which illustrates the sequences of HIV-1 and other retroviral proteinases. These residues include Pro 1, Ile 3, and Leu 5 of the N-terminal strand, and Cys 95, Leu 97 and Phe 99 of the C-terminal strand that comprise one side of the central four-stranded β-pleated sheet; this packs onto Leu24 and other residues within the subunit core, so stabilising the dimer. Hydrophobic residues of the helix such as Leu 89, Leu 90 and Ile 93 also contribute to the hydrophobic core. All of these residues tend to be invariant or conservatively varied amongst the

a)

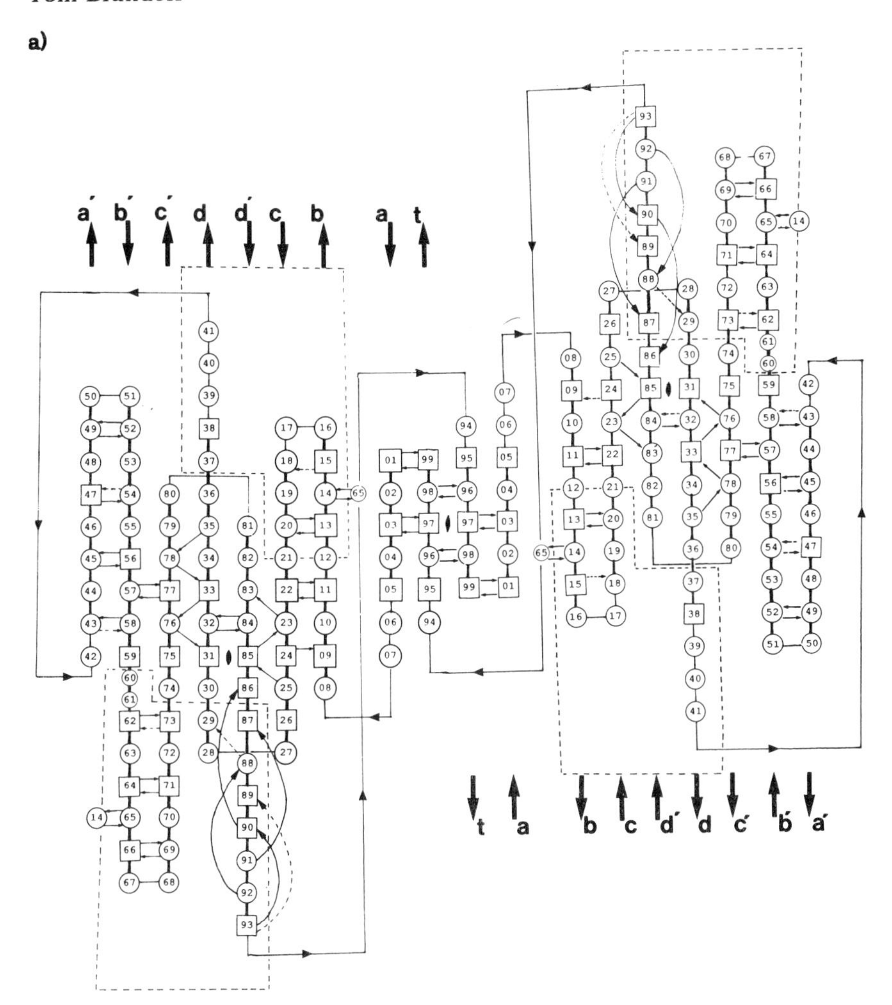

Figure 5.
Schematic diagrams of the super-secondary structures of (a) the HIV-1 proteinase and (b) the aspartic proteinase, endothiapepsin, showing the arrangment of β-strands, α-helices, and hydrogen bonds. Well defined hydrogen bonds are shown by an arrow from the donor to the acceptor; weaker hydrogen bonds with distances less than 3.9Å are shown with a dotted line. Hydrogen bonds to side chains are not shown. The β-strands are labelled according to the scheme of Blundell et al. (1985) to emphasise the pseudo-symmetry between the motifs and domains; for example, strands *b* and *b'* are related by a pseudo two-fold axis perpendicular to the paper indicated by ⬮. Residues with sidechains that are inaccessible to solvent are indicated by squares and those accessible to solvent are indicated by circles. Dotted lines include strands that are folded onto the top of the main sheet and are hydrogen-bonded residues 14 and 65.

Figure 5b.

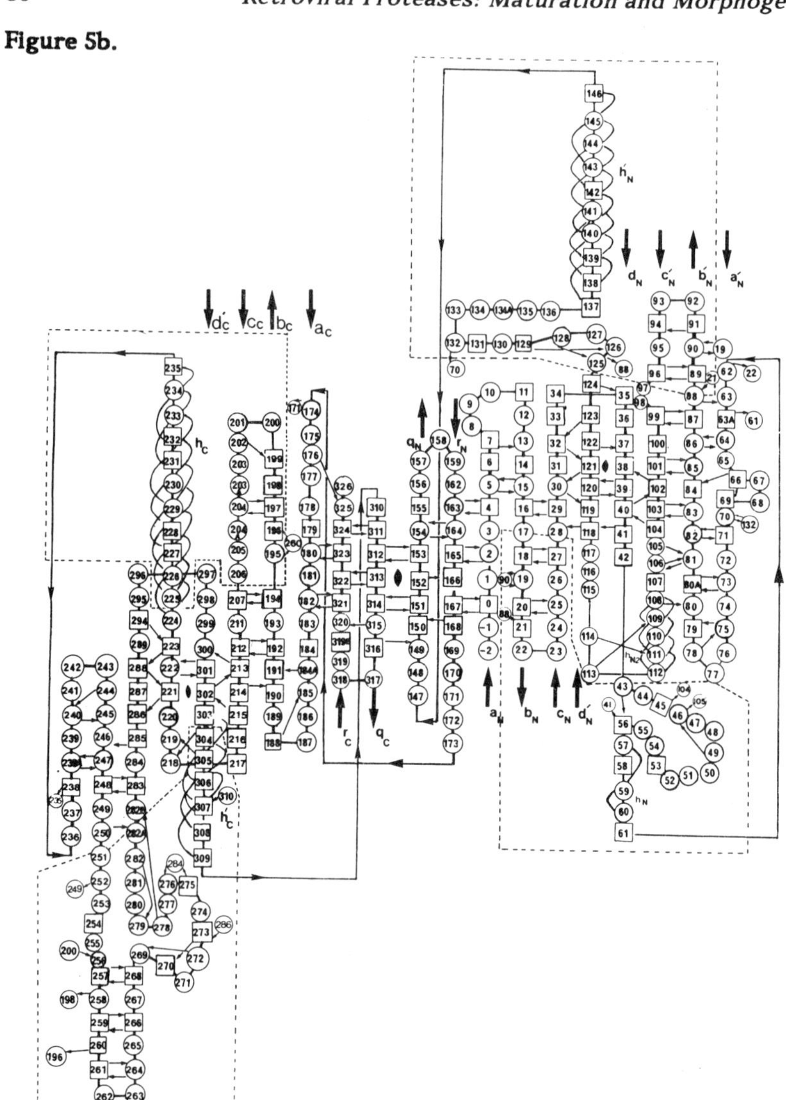

retroviral proteinases, giving support to this interpretation of the structure. The only truly buried polar residue apart from Thr 26 (see below) is Thr 31; however, this is also hydrogen bonded to mainchain NH and CO functions in a fashion that is often found to be important to protein structure. Thr 31 is usually retained as threonine in retroviral proteinases although it is occasionally conservatively varied to serine.

One of the significant conservations in the consensus sequences of the retroviral and pepsin-like aspartic proteinases is the sequence hydrophobic-hydrophobic-glycine, for example Ile 84-Ile 85-Gly 86 found in HIV-1 proteinase. However, the retroviral proteinases differ in the presence of an invariant arginine in the following position, for example Arg 87. Such a basic residue is never found in the pepsins; it is always either a hydrophobic residue or an aspartic acid. The crystal structure of HIV-1 proteinase shows that as in RSV-proteinase (Miller *et al.* 1989) this conserved arginine participates in an intersubunit cluster of ionic and hydrogen bonding interactions involving the sidechains of the conservatively varied Asp 29 and Arg 8 and the CO function of Gln 7.

The catalytic residues.

The conserved active site residues, Asp 25-Thr 26-Gly 27, form a symmetrical and highly hydrogen-bonded arrangement virtually identical to that described for pepsin-like aspartic proteinases (James and Sielecki, 1983; Pearl and Blundell, 1984)). This includes the two threonines, which are inaccessible to solvent and hydrogen- bonded so that the Oγ hydrogen bonds to mainchain NH and CO functions of the other subunit in a fireman's grip. The two aspartates lie approximately planar with their inner carboxylate oxygens hydrogen bonded to the NH functions of Gly 27 and within hydrogen-bonding distance of each other. There is density for a water molecule bound equally to the two carboxylates as found in the pepsin-like enzymes.

Comparisons with 3-D structures of aspartic proteinases.

Using a least squares algorithm (Sutcliffe *et al.*, 1987)), we have compared the structure determined here for the HIV-1 proteinase with models proposed earlier for the HIV-1 proteinase. The model of Pearl and Taylor (1987) gave root mean square differences for all Cα-atoms of approximately 4.0Å; while those based on RSV proteinase (for example Weber *et al.*, 1989) varied from 3.2 to 4.0Å. Most of these errors were due to misalignments of the sequences in loop regions and the models gave remarkably good estimates of the active site regions. We also compared the structure to those defined by X-ray analysis. For the RSV proteinase (Miller *et al.* 1989) 67 residues of the subunit could be aligned with topologically equivalent residues in the structure reported here with a root mean square difference of 1.32Å. Strictly rigid group superposition of the HIV-1 proteinase structure on pepsin from hexagonal crystals (Cooper and Blundell, 1989) gave only 18 residues simultaneously common between each of the two domains of the pepsin and the subunit of the retroviral proteinase with a root mean square difference of 0.86Å with the N- domain and 0.72Å with the C-domain. However, as proposed earlier (Pearl and Taylor, 1987; Blundell *et al.* 1988) nearly all the strands in the pepsins have equivalents in the retroviral proteinases.

Specificity sites and substrate binding.

The close similarity of the three-dimensional structures of the dimeric retroviral proteinases and the monomeric pepsins allows some suggestions concerning the mode of interaction of the enzymes with the substrate. Much evidence concerning this has been inferred from X-ray studies of aspartic proteinase-transition state isostere complexes (Bott *et al.*, 1982; James *et al.*,1982; Blundell *et al.* 1987). These show that the substrate probably binds pseudo-symmetrically in the active site cleft using topologically equivalent hydrogen bonding functions and specificity pockets on each side of the scissile bond. In HIV-1 proteinase this implies by analogy that the carbonyls of Gly 27 hydrogen-bond to the substrate NH functions of P1 and P2' as suggested by Blundell and Pearl (1989) and by Wlodawer and coworkers (1989). The conservatively varied and exposed Leu 23, Ala 28, Val 82 and Ile 84 form the large S1/S1' specificity subsites. By analogy with the aspartic proteinases where P3 binds through its mainchain to the sidechain Oγ and the mainchain NH of Thr 219 (pepsin numbering), this interaction would be mediated by Asp 29 in HIV- proteinase. The mainchain NH is available for hydrogen bonding and the conservatively varied Asp sidechain might bind the NH function of P3. This gives further support to the importance of the cluster of charged residues involving Asp 29; it is possible that the charged residue Arg 87 contributes to the correct orientation of the carboxylate group. The pocket, S3, involves the loop between strands *a* and *b*.

The perfectly symmetrical structure of the retroviral proteinase structure lends credence to the suggestion (Pearl and Blundell, 1984) that the apparently differing pK_a's of the two aspartates in aspartic proteinases are more a function of their close coexistence rather than their differing environments. Nevertheless the symmetrical dimer must lose its perfect symmetry during catalysis as the substrate has only an approximate dyad. This is, however, not necessary for inhibitors so that symmetrical inhibitors may have a major advantage in high affinity binding when compared to an analogous substrate.

Modelling other retroviral proteinases.

Figure 6 shows the alignment of several retroviral proteinase sequences. The alignment was obtained by first comparing the 3-dimensional structures of RSV and HIV-1 proteinases by direct superposition and by a program COMPARER (Sali and Blundell, 1990) to obtain topological equivalences and then by using this alignment to guide the alignment of other retroviral proteinase sequences. We have used these alignments to model the 3-dimensional structures of representative retroviral proteinases from several classes in order to investigate the effect of sequence variation on 3-dimensional structure and specificity. We first considered HIV-2, SIV_{agm} and SIV_{mac} proteinases as well as the more distantly related enzyme from FIV. Here we shall present a brief discussion of our studies of the HIV-2 and FIV proteinases.

```
RSV      LAMTME-----HKDRPLVRVILTNTGSHPVKQRSVYITALLDSGADITII
FIV      VNYNKVGTTTTLEKRPEILIFVN-----------GYPIKFLLDTGADITIL
SIV-agm  FELPL--------WRRPIKTVYIE-----------GVPIKALLDTGADDTII
HIV-2    PQFSL--------WKRPVVTAYIE-----------GQPVEVLLDTGADDSIV
HIV-1    PQITL--------WQRPLVTIKIG-----------GQLKEALLDTGADDTVL
         1                10                      20        30


RSV      SEEDWPT--DWPVMEAANPQIHGIGGGIPMRKSRDMIELGVINRDGSLER
FIV      NRRDFQVKNSIENGRQN---MIGVGGGKRGTNY-INVHLEIRDEN-YKTQ
SIV-agm  KENDLQ--LSGPWRPKI---IGGIGGGLNVKEY-NDREVKIE------DK
HIV-2    AGIELG----NNYSPKI---VGGIGGFINTKEY-KNVEIEVL------NK
HIV-1    EEMSLP----GRWKPKM---IGGIGGFIKVRQY-DQILIEIC------GH
                   40              50          60


RSV      PLLLFPAVAM---VRGSILGRDCLQGLGLRLTNL
FIV      CIFGNVCVLEDNSLIQPLLGRDNMIKFNIRLVM
SIV-agm  ILRGTILLGA---TPINIIGRNLLAPAVPRLVM
HIV-2    KVRATIMTGD---TPINIFGRNILTALGMSLNL
HIV-1    KAIGTVLVGP---TPVNIIGRNLLTQIGCTLNF
         70            80           90
```

Figure 6.
Alignment of amino acid sequences of some retroviral proteinases.

The alignment of the sequences of proteinases from HIV-1 and HIV-2 was straightforward as they exhibit approximately 85% residue identity. A model for the HIV-2 proteinase was constructed from the structure of the HIV-1 enzyme using the automated procedure of the COMPOSER program (Blundell *et al.* 1988). This was followed by manual optimization of amino acid packing and geometry using the program FRODO (Jones, 1978) on an Evans and Sutherland PS390 graphics system. These modelling studies showed that the sequence of the HIV-2 enzyme can adopt a tertiary structure that is very similar to the HIV-1 enzyme. It can have an identical arrangement of catalytic residues and very similar specificity pockets. However, the substitution of a valine for an isoleucine in HIV-2 at position 82 in the S1/S1' specificity pockets is expected to lead to a preference for a smaller hydrophobic amino acid at P1 and P1' in the substrate.

Modelling the FIV proteinase was less straightforward, mainly because there are some ambiguities in the alignment of the sequence with that of HIV-1 proteinase. Insertions in the FIV sequence relative to that of the HIV-1 enzyme occur in external loop regions. Alignment of the sequences of the HIV-1 proteinase with those of aspartic proteinases from mammalian and fungal sources performed using the program

COMPARER (Sali and Blundell, 1990; Z. Zhu and A. Sali, unpublished results) indicated that insertions similar to those found in FIV relative to HIV-1 are also found in the N- terminal domains of pepsins. Specifically, the insertion at the end of the *d*-strand could be built as a loose loop packing against a slightly longer insertion at the *c*', *d*'-loop. Similar insertions are found in a number of the pepsins for which X-ray structural information is available. This model, while preliminary, does suggest a basic specificity difference between the HIV-1 and FIV enzymes. The residues of the insertion at the wide loop between the *c*' and *d*' strands may contribute to the S1 and S3 (S1' and S3') specificity pockets relative to HIV-1. Such differences are likely to be of importance when trying to assess the suitability of the FIV retrovirus as a suitable animal model for the human retrovirus.

ACKNOWLEDGEMENTS

We thank Dr Simon Campbell, Dr Colin Greengrass, Graham Rickett, Dr Jon Cooper, Dr Lynn Sibanda and Andrej Sali for their help and advice. We thank the UK AIDS Directed Programme (AW, LHP), University of Helsinki and Osk. Huttunen Foundation (RL), Cray Reasearch (AH) and the Royal Society (ZZ) for financial support.

REFERENCES

Blundell, T.L., Jenkins, J.A., Pearl, L.H., Sewell, B.T. and Pedersen, V. in 'Aspartic Proteinases and their inhibitors', (Kostka, V.,ed.), pp. 151-161, Walter de Gruyter, Berlin (1985).

Blundell, T.L., Cooper, J., Foundling, S.I., Jones, D.M., Atrash, B and Szelke, M. *Biochemistry*, **26**, 5585-5590 (1987).

Blundell, T.L., Carney, D., Gardner, S., Hayes, F., Howlin, B., Hubbard, T., Overington, J., Singh, D.A., Sibanda, B.L. and Sutcliffe, M. *Eur. J. Biochem.*, **172**, 513-520 (1988).

Blundell, T.L. and Pearl, L.H. *Nature*, **337**, 596-597 (1989).

Bott, R.,Subrmnian, E. and Davies, D.R. *Biochemistry*, **21**, 6956-6962 (1982).

Cooper, J. and Blundell, T.L. (1989) Deposited in the Brookhaven Data Base.

Darke, P.L., Leu, C.-T., Davis, L.J., Heimbach, J.C., Diehl, R.E., Hill, W.S., Dixon, R.A.F. and Sigal, I.S. *J. Biol. Chem.*, **264**, 2307-2312 (1989).

Darke, P.L., Nutt, R.F., Brady, S.F., Garsky, V.M., Ciccarone, T.M., Leu, C.-T., Lumma, P.K., Freidinger, R.M., Veber, D.F. and Sigal, I.S. *Biochem. Biophys. Res. Comm.*, **156**, 297-301 (1988).

Franke, A.E. *Eur. Patent No.* 0147178 (1985).

Haneef, I., Moss, D.S., Stanford, M.J. and Borkakoti, N. *Acta Cryst.*, **A41**, 426-433 (1985).

Hobart, P.M., Lee, S.E. and Geoghegan, K.F. in preparation.

James, M.N.G., Sielecki, A.R., Salituro, F., Rich, D.H. and Hofmann, T. *Proc. Nat. Acad. Sci. USA*, **79**, 6137-6142 (1982).

James, M.N.G. and Sielecki, A. *J. Mol. Biol.*, **163**, 299-361 (1983).

Jones, T.A. *J. Appl. Cryst.*, **11**, 268-272 (1978).

Lapatto, R., Blundell, T.L., Hemmings, A., Overington, J., Wilderspin, A.W., Wood, S.P., Merson, J., Whittle, P., Danley, D.E., Geoghegan, K.F., Hawrylik, S., Lee, S.E., Scheld, K. and Hobart, P.M. *Nature*, **342**, 299–302, (1989)

McKeever, B.M., Navia, M.A., Fitzgerald, P.M.D., Springer, J.P., Leu, C.-T., Heimbach, J.C., Herber, W.K., Sigal, I.S. and Darke, P.L. *J.Biol. Chem.*, **264**, 1919 (1989).

Miller, M., Jaskólski, M., Rao, J.K.M., Leis, J. and Wlodawer, A. *Nature*, **337**, 576–579 (1989).

Navia, M.A., Fitzgerald, P.M.D., McKeever, B.M., Leu, C.-T., Heimbach, J.C., Herber, W.K., Sigal, I.S., Darke, P.L. and Springer, J.P. *Nature*, **337**, 615–620 (1989).

Pearl, L.H. and Blundell, T.L. *FEBS Lett.*, **174**, 96–101 (1984).

Pearl, L.H.and Taylor, W.R. *Nature*, **329**, 351–354 (1987).

Ratner, L., Haseltine, W., Patarca, R., Livak, K.J., Starcich, B., Josephs, S.F., Doran, E.R., Rafalski, J.A., Whitehorn, E.A., Baumeister, K., Ivanoff, L., Petteway, S.R., Pearson, M.L., Lautenberger, J.A., Papas, T.S., Ghrayeb, J., Chang, N.T., Gallo, R.C. and Wong Staal, F. *Nature*, **313**, 277–284, (1985).

Sutcliffe, M.J., Haneef, I., Carney, D. and Blundell, T.L. *Prot. Eng.*, **1**, 377–384 (1987).

Tang J., James, M.N.G., Hsu, I.-N., Jenkins, J.A. and Blundell, T.L. *Nature*, **271**, 618–621 (1978).

Toh H., Ono, M., Saigo, K. and Miyata, T. *Nature*, **315**, 691 (1985).

Weber, I.T., Miller, M., Jaskólski, M., Leis, J., Skalka, A.M. and Wlodawer, A. *Science*, **243**, 928–932 (1989).

Wlodawer, A., Miller, M., Jaskólski, M., Sathyanarayana, B.K., Baldwin, E., Weber, I.T., Selk, L.M., Clawson, L., Schneider, J. and Kent, S.B.H. *Science*, **246**, 616–621 (1989)

10
X-Ray Analysis of HIV-1 Protease and Its Complexes with Inhibitors

Maria Miller, Amy L. Swain, Mariusz Jaskólski, Bangalore K. Sathyanarayana, Garland R. Marshall, Daniel Rich, Stephen B. H. Kent and Alexander Wlodawer

The HIV-1 protease is essential for the replication of infective virus (Kohl *et al.*, 1988), and is therefore an attractive target for the design of specific inhibitors as potential antiviral therapeutics. The rational design of drugs that can inhibit the action of viral proteases depends on obtaining accurate structures of these enzymes. Chemically synthesized HIV-1 protease (Schneider and Kent, 1988; Merrifield, 1963; Kent, 1988; Tam *et al.*, 1986) was used to solve crystal structures of unliganded enzyme and of complexes with two peptide inhibitors. The amino acid sequence of the enzyme, corresponding to the SF2 isolate is shown on Figure 1.

(H)-P^{1} Q I T L W Q R P L^{10} V T I [R] I G G Q L K^{20}

E^{21} A L L D T G A D D^{30} T V L E E M N L P G^{40}

[K]^{41} W K P K M I G G I^{50} G G F I K V R Q Y D^{60}

Q^{61} I [P] [V] E I [Aba] G H K^{70} A I G T V L V G P T^{80}

P^{81} V N I I G R N L L^{90} T Q I G [Aba] T L N F^{99}-(OH)

Figure 1.
Amino acid sequence of [Aba^{67,95}]HIV-1 protease (SF2 isolate). The residues that differ from the NY5 isolate (6) are boxed (replacements are R, K; K, R; P, L; and V,I). L-α-Amino-*n*-butyric acid (Aba) was used in place of the two Cys residues.

In this enzymatically active protein, the cysteines were replaced by α-amino-*n*-butyric acid, a nongenetically coded amino acid. The resulting enzyme preparation was homogeneous by reverse phase HPLC and SDS-PAGE, and had the correct covalent structure, as determined by mass spectrometric peptide mapping. The synthetic enzyme had the specific proteolytic activity characteristic of the HIV-1 protease, and had a turnover number comparable to that reported by Darke *et al.* (1989, 1988) for the enzyme derived from bacterial expression.

CRYSTAL STRUCTURE OF HIV-1 PR

Single crystals of HIV-1 PR were grown by a modification of the method described by McKeever *et al.*, (1989) and were isomorphous with those obtained by them (tetragonal space group $P4_12_12$, unit-cell parameters **a** = 50.24Å, **c** = 106.56Å). The number of reflections measured to 2.8Å resolution was 26,037, which reduced to 3225 unique data, of which 2614 were considered observed ($I>1.5\sigma(I)$).

The crystal structure (Wlodawer, *et al.*, 1989) was solved first by molecular replacement and then confirmed by the multiple isomorphous replacement (MIR) method. The molecular replacement method used the HIV-1 protease model constructed by Weber *et al.* (1989) based on the known structure of the RSV PR (Miller *et al.*, 1989). The final *R*-factor after the refinement is 0.184 with acceptable geometry (deviations from ideal bond lengths of 0.026Å), and with average B = 16.9$Å^2$.

The general topology of the HIV-1 PR molecule (Figs. 2,6a) is similar to that of a single domain in pepsin-like aspartic proteases (Blundell et al., 1985). The main difference is in the interface region where the two HIV-1 PR monomers have significantly shorter N- and C-terminal strands which interact to produce a dimer whose structure and properties are similar to those of the two-domain, pepsin-like enzymes. The N-terminal β-strand *a* (residues 1-4) forms the outer part of the interface β-sheet. The *b* β-strand (9-15) continues through a turn into the *c* β-chain which terminates at the active site triplet (25-27). Following the active-site loop is the *d* β chain with residues 30-35. In pepsin-like proteases, chain *d* is followed by the *h* helix which in RSV PR is reduced to a short and distorted segment. In HIV-1 PR this fragment is even more distorted and has the form of a broad loop (36-42). The second half of the molecule has a topology related to that described above by an approximate intramolecular two-fold axis (corresponding substructures indicated by primed labels). Residues 43-49 form the *a'* β-strand which, as in pepsin-like proteases, belongs to the flap. The other strand in the flap (52-58) forms a part of the long *b'* β-chain (52-66). The flap is most probably stabilized by a pair of hydrogen bonds between main-chain atoms of residue 51, located at the tip of the flap, and its symmetry mate in the crystallographic dimer. The *c'* β-chain comprises residues 69-78 and after a loop at 79-82 continues as chain *d'* (83-85) which leads directly to the well-defined helix *h'* (86-94). The hydrogen bonding pattern within this helix is intermediate between the α- helix and the 3_{10} helix. Helix *h'* is followed by a straight C-terminal β strand (95-99), which can be designated as *q*, in a way analogous to RSV PR. In contrast, pepsin-like aspartic proteases have a double-stranded b sheet here. This C-terminal region forms the inner part of the intra-dimer interface. Four of the b strands in the molecular core are organized into a Ψ-shaped sheet characteristic of all aspartic proteases. One of the Ψ letters comprises chains *c*(23-25), *d*, *d'* and the other is made up of *c'*(76- 78), *d'*, *d*.

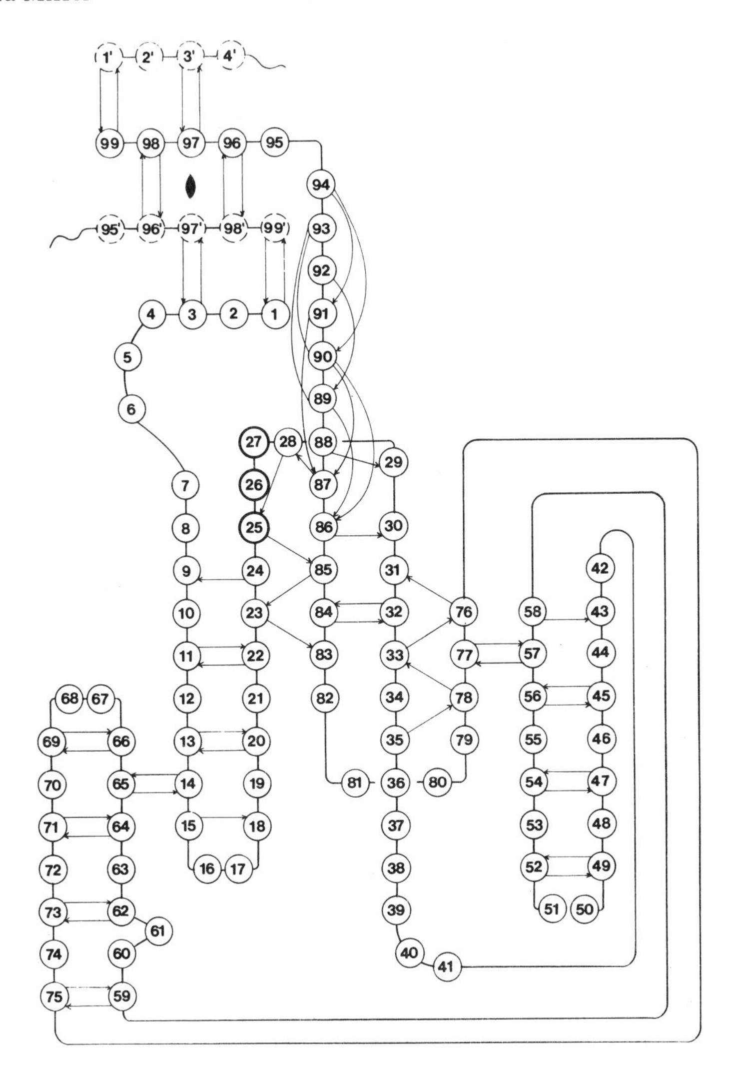

Figure 2.
Diagram of the secondary structure of the HIV-1 protease. Main-chain NH...O hydrogen bonds are indicated by arrows (---->). The active site (Asp-Thr-Gly) is marked by thicker circles, and the broken circles correspond to residues from the symmetry-related molecule in the dimer interface.

The active-site triad (Asp25-Thr26-Gly27) is located in a loop whose structure is stabilized by a network of hydrogen bonds which is characteristic of aspartic proteases. The symmetry-related Asp carboxylic groups from the two loops composing the active site are nearly coplanar, and show a close contact involving their OD1 atoms. Additionally, the two active-site loops are linked by the 'fireman's grip' in which each threonine (the middle residue in the active-site triplet) accepts a hydrogen bond from the amide group of the threonine in the other loop and donates a hydrogen bond to the carbonyl O atom in the residue preceding the catalytic triad on the other strand. The structure of each individual loop is reinforced by a hydrogen bond between OD1 of the aspartate and the NH group of the glycine, the last residue in the triplet. In all pepsin-like enzymes the residue immediately following the active-site triplet has a hydroxyl group (the only exception is one domain of human renin which has alanine in this position), which usually forms a hydrogen bond with the side chain of the catalytic Asp on the same strand. This is in contrast with retroviral proteases which invariably have Ala in this position, so that no such hydrogen bonding is possible.

The final model of the HIV-1 PR was compared with the structure of RSV PR, which has been refined to $R = 0.144$ at 2Å resolution (Jaskólski *et al.*, unpublished). The agreement is very close, with an rms deviation of 1.5Å for 86 common Cα atoms in one subunit. With minor exceptions, the corresponding Cα 's follow closely the sequence alignment proposed previously (Weber *et al.*, 1989). Detailed comparisons, however, showed that deletions of several residues from the RSV PR sequence, in addition to those previously proposed, were necessary to properly align it with the HIV-1 PR. They include residues 7, 8, 48, 49 and 93 (RSV numbering). Since the structure of the flap region was not seen in the RSV protease (Miller *et al.*, 1989), no correction in the alignment is possible for that region.

Following the previous comparisons of the structure of RSV PR and pepsin-like proteases (Erickson *et al.*, unpublished), the present structure of HIV-1 PR has been compared with both domains of rhizopuspepsin (Suguna *et al.*, 1987). Contrary to previous observations (Navia *et al.*, 1989), we found no difference between the fit to the C- and N-domains. For the N- domain of rhizopuspepsin, we could superimpose 57 Cα pairs (over half of all residues in HIV-1 PR) to an rms deviation of 1.4Å, while 56 Cα pairs in the C-domain superimposed within the same limits. It must be stressed that different elements of the secondary structure in each domain are responsible for some of these alignments.

Comparisons of all atoms in the active sites (the catalytic triad flanked by four residues on each side), showed that 126 atom pairs superimposed with an rms deviation of 0.57Å between HIV-1 PR and RSV PR, while the optimum superposition between HIV-1 PR and rhizopuspepsin included 88 pairs deviating by 0.59A .

The structure described here was recently confirmed for cloned material (Lappato, *et al.*, 1989). It differs from that reported earlier (Navia *et al.*, 1989) in several important details. Residues 86 to 94 form a helix in close agreement with the model structure derived from the RSV protease (Weber *et al.*, 1989) and the sequence and structural similarities of viral and pepsin-like proteases (Erickson *et al.*, unpublished). Residues 1 to 5 which were previously reported to be disordered, are part of a β-sheet composed of alternate strands from the NH - and COOH- termini of each subunit in the dimer.

In conclusion our studies confirm conserved folding in retroviral proteases, and suggest that only intermolecular catalysis is responsible for the release of the protease in both RSV and HIV-1. In addition, precise information of the specific interactions of the termini should be useful in the design of inhibitors of dimer formation.

STRUCTURE OF THE ENZYME-INHIBITOR COMPLEXES

While a structure of an enzyme with no active-site ligand conveys only limited information about the enzymatic process and inhibitor design, much more detailed information can be gleaned from studies of enzyme complexed with substrate-derived inhibitors. This approach has been extensively applied in studies of cell-derived aspartic proteases, particularly with the aim of producing inhibitors of human renin which could act as anti-hypertensive drugs. While no structures of renin-inhibitor complexes have been published to date (Sielecki *et al.*, 1989), extensive investigation of inhibitor complexes with endothiapepsin (Sali *et al.*, 1989), rhizopuspepsin (Suguna *et al.*, 1987), and penicillopepsin (James and Sielecki, 1987) have been reported. For the HIV-1 PR, the observed similarity of the active sites of cellular and retroviral proteases led to the modeling of the interaction of inhibitors and substrates with the enzyme (Weber *et al.*, 1989). While the resulting models were reasonable, it was clear that many details could not be unambiguously determined in the absence of experimental data. It was also clear that the available crystal form of the HIV-1 PR was not suitable for inhibitor studies, due to the presence of only one subunit (half of the molecule) in the asymmetric unit and to the limited resolution of the measurable diffraction data (Navia *et al.*, 1989; Wlodawer *et al.*, 1989). For these reasons, we sought to cocrystallize HIV-1 PR with a substrate-based inhibitor. The hexapeptide substrate:

Ac-Thr-Ile-Met-Met-Gln-Arg.amide

with K_m (1.4mM) was chosen as a candidate for further modification. The isosteric amino acid, norleucine (Nle), in which the sulfur of the methionine side chain is replaced by a methylene group, was used for synthetic simplification and was shown to give a substrate:

Ac-Thr-Ile-Nle-Nle-Gln-Arg.amide

with comparable affinity. In an approach analogous to that of Szelke et al. (1982) for human renin, an inhibitor of HIV-1 PR based on this substrate, compound MVT-101, with the sequence:

N-acetyl-Thr-Ile-Nle-Ψ[CH_2-NH]-Nle-Gln-Arg.amide

was prepared, where the scissile peptide bond has been replaced by a reduced analog. The K_i for MVT-101 was determined to be 0.78 mM (M.V. Toth and G.R. Marshall, unpublished results). For inhibitor binding, enzyme (stored in phosphate buffer, pH 7, in 20% glycerol in -20°C) was concentrated to 6 mg/ml using Centricon-10 microconcentrators, while simultaneously exchanging the buffer for 20 mM sodium acetate, pH 5.4.

5 mg of MVT-101 was dissolved in 100 μl DMSO and mixed with protein to yield 10-fold molar excess of the inhibitor (20-fold over homodimeric molecule). Crystals grew at room temperature in hanging drops from 60% ammonium sulfate. They appeared within a few days as thin rods with maximum dimensions of 0.3 x 0.12 x 0.06 mm. The crystals belonged to an orthorhombic space group, $P2_12_12_1$ with the unit cell lengths **a**= 51.7 Å, **b** = 59.2 Å, **c** = 62.45 Å, and the asymmetric unit contained two identical 99-residue polypeptide chains (forming a homodimeric enzyme molecule), and one inhibitor molecule. 2.0 Å data were collected from two crystals. The present model which includes 70 water molecules, is characterized by an *R* factor of 17.6% for 7813 reflections between 10 and 2.25 Å, with the deviations from ideality of 0.019 Å for bonds, 0.014 Å for the planes, and 0.22 Å for the chiral volumes. Further refinement including higher resolution data is in progress.

The crystal structure was solved by molecular replacement using the previously solved structure of the native HIV-1 PR (Wlodawer *et al.*, 1989) as a starting model. The $|F_o-F_c| \alpha_c$ map based on the phases from the preliminary refinement showed density in the active site which could be recognized as corresponding to the inhibitor. A model of the hexapeptide with a reduced peptide bond could be fitted very easily, with the polarity indicated primarily by the bulky side chain of Arg206. The positions of all six residues of the inhibitor are well defined by the electron density (Fig. 3). The inhibitor binds to the active site of the protease in an extended conformation with backbone angles shown in Table I. The reduced peptide bond between the two norleucine residues deviates from planarity by 38° - a value similar to that reported for the reduced peptide inhibitor of rhizopuspepsin (Suguna *et al.*, 1987). The inhibitor molecule makes extensive interactions with the enzyme at the interface

Table I
The backbone dihedral angles (degrees) of the substrate-derived inhibitor in the enzyme-MVT-101 complex. These angles correspond to an extended conformation of the polypeptide.

Residue	φ	ψ	ω
Thr 201		125	-176
Ile 202	-121	92	-178
Nle 203	-92	72	142
Nle 204[a]	-92	154	-179
Gln 205	-151	123	178
Arg 205	-81	129	

[a] Reduced peptide bond.

between the protein subunits. It is more than 80% excluded from contact with the surrounding solvent by the protein molecule, with only the N- and C-terminal amino acids partially exposed.

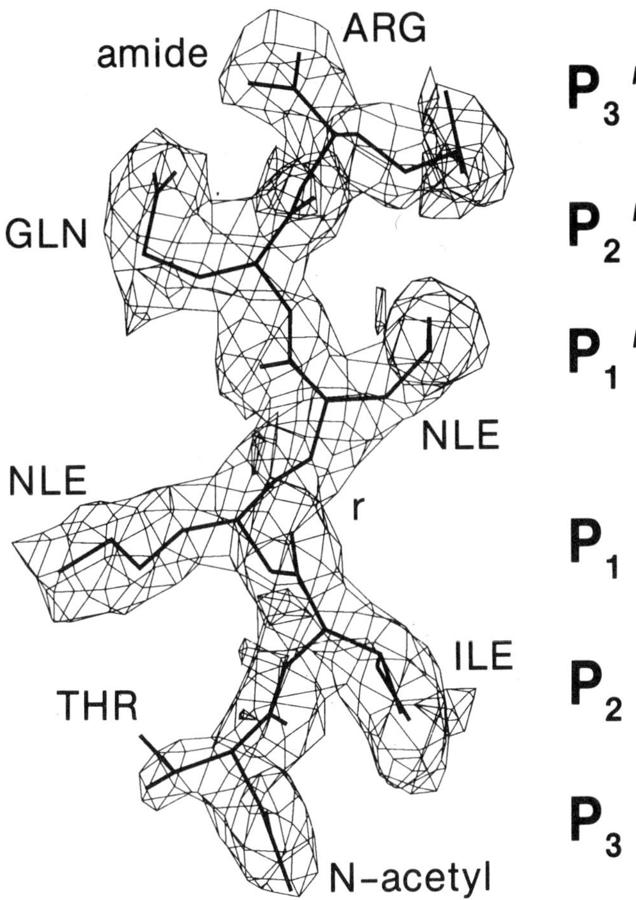

Figure 3.
A view of the electron density and of the final atomic model of the inhibitor MVT-101. This $|2F_O-F_C|\alpha_C$ electron density map was calculated after refinement and was contoured at 0.8σ level. All atoms belonging to the inhibitor molecule, with the exception of the side chain CG of Thr 201, are in the density.

A schematic representation of hydrogen bonds between the inhibitor and its environment in the active site cleft is shown in Figure 4. There are six direct inhibitor-enzyme hydrogen bonds between the main chain oxygen and nitrogen atoms. These involve Asp29 N, Gly27 O and Gly48 O from both monomers. The flaps (Ile50 N) bind through a water bridge to carbonyls of inhibitor residues in the P2 and P1' subsites. The side chain carboxyl of Asp29 forms a long hydrogen bond with a nitrogen of N-acetyl terminus, while the carboxyl of Asp29' interacts with the C-terminus amide group. Two water molecules are hydrogen bonded to the amido carbonyls of the end-groups of the reduced peptide. As predicted by model building based on the analogy with cellular proteases (Weber *et al.*, 1989; Lapatto *et al.*, 1989), specific hydrogen bonded interactions with the main chain of the inhibitor in P1/P2 and P1'/P2' are made by the carbonyl oxygens of Gly27 and Gly27'. This interaction necessitated the rotation of the planes of the Gly27 and Gly27' peptide groups by almost 90° compared to the native structure. Extensive van der Waals contacts with the enzyme residues define the binding pockets of the complex and are listed in the Table II. Side chains P2, P1, P1', P2' are located in very well defined hydrophobic pockets; on the contrary, Arg206 is held at S3' by ionic interactions with the carboxylate of Asp29', while Thr201 at S3 does not seem to interact strongly with the enzyme and its side chain may be disordered, as indicated by the lack of well defined electron density (Fig. 3).

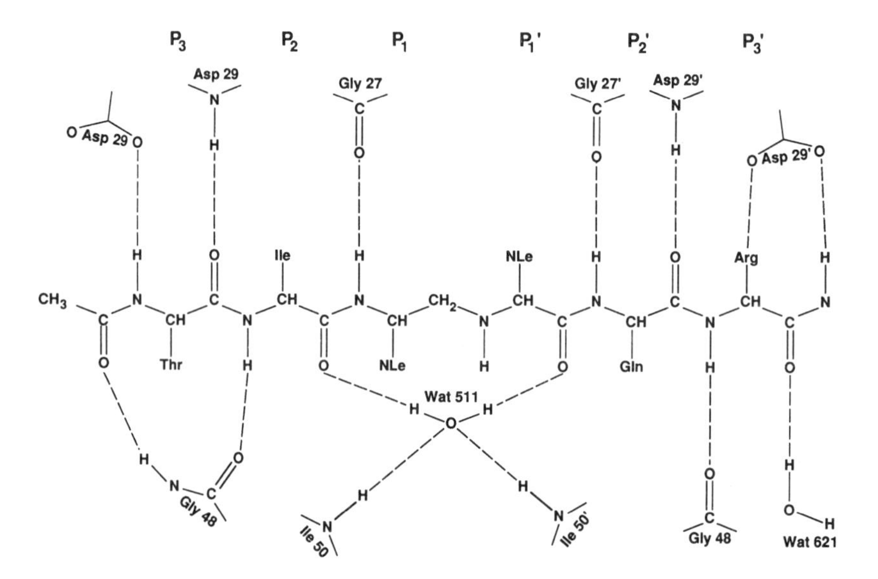

Figure 4.
Schematic representation of the hydrogen bond interactions between the inhibitor and the protein.

Table II

Contacts between HIV-1-protease dimer and the substrate-derived inhibitor, MVT-101. The binding pocket is defined as those residues of the protein which are within 4.2Å radius of the inhibitor (8). Underlined residues have moved by more than 2Å (at Cα) as a result of inhibitor binding.

Subsite (Inhibitor)		Binding pockets (enzyme)			
P3	Thr201	S3:	Arg8',	Asp29,	Gly48
P2	Ile202	S2:	Ala28, Ile84	Ile47,	Ile50',
P1	Nle203	S1:	Leu23', Pro81',	Asp25', Val82'	Ile50, Ile84'
P1'	Nle204	S1':	Leu23, Ile50', Ile84	Gly27, Pro81,	Asp25, Val82,
P2'	Gln205	S2':	Val32', Ile50	Ile47',	Gly48',
P3'	Arg206	S3':	Arg8, Gly48',	Gly27', Val82	Asp29',

Binding of the inhibitor introduces substantial conformational changes to the enzyme (Fig. 5). The overall movement of the subunits can be described as a hinge motion by 1.7°, with the hinge axis located in the intersubunit β-sheet interface, and the rotation axis perpendicular to the view shown in Figure 5. This motion slightly tightens the cavity of the active site, but as expected from model building (Weber *et al.*, 1989), the major rearrangement of that area involves very large motion of the flap regions. This motion is significant for as much as a quarter of each subunit, since all Cα atoms for residues 34-57 in either flap shift by over 1Å, with the movement as large as 7Å for the tips of the flaps (around residue 50). The movement of both flaps is grossly symmetric, with maximum deviation between equivalent Cα atoms in each of less than 1Å. The major asymmetric element lies in the tip of the flap and is introduced by the flip of a peptide bond between residues 50' and 51' compared to their counterparts in the other molecule. As a result, only one hydrogen bond (N51...O50') connects these polypeptide chains, while N51' and O50 hydrogen bond to water molecules only.

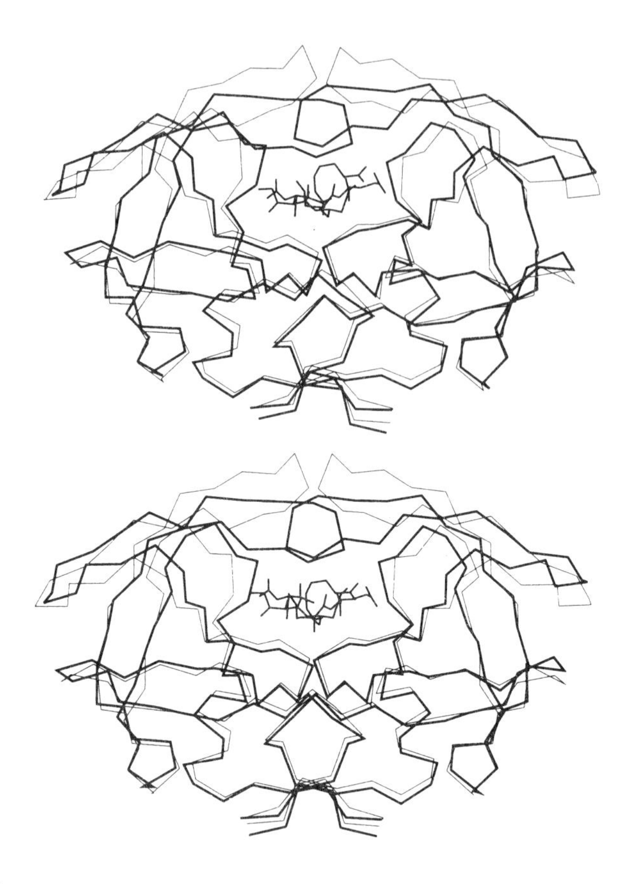

Figure 5.
Stereo tracing of the superimposed Cα backbones of the native (thin lines) and inhibited (thick lines) protease dimers, with the inhibitor as reported here marked in medium lines. The large movements of the flap regions, as well as of the loop containing residues 79–82, can be clearly seen.

There is an interesting topological change between the arrangements of the tips of the flaps in the native and inhibited HIV-1 PR which can be easily seen in Figure 6. While the flap of the bottom subunit in the native enzyme is above the one from the top subunit, the opposite is true for the inhibited enzyme. This behavior can be explained if the flaps rearrange themselves after opening which allows the inhibitor to enter the binding cleft. It is also possible that the arrangement of the flaps in the native enzyme is an artefact of crystal packing, since the tips of the flaps are involved in intermolecular contacts (Navia *et al.*, 1989; Wlodawer *et al.*, 1989).

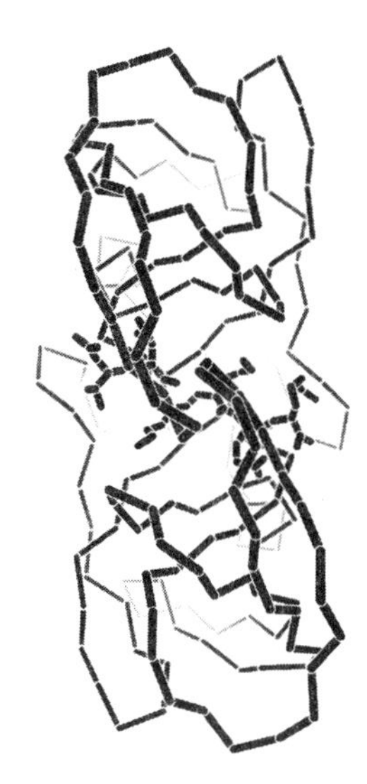

Figure 6.
$C\alpha$ backbone of HIV-1 protease dimer (a) for nonliganded enzyme (b) with the MVT-101 hexapeptide in the active site in the view showing rearrangement of the flaps.

Another region which involves substantial movement of the protein backbone on binding of the inhibitor involves residues 77-82 in both subunits. The maximum shifts in the $C\alpha$ positions in that region exceed 2Å, and the direction of movement also decreases the size of the active site cavity. The driving force for this motion is a hydrophobic interaction of Pro81 (81') and Val82 (82') with norleucines 203 and 204 (inhibitor subsites P1 and P1'); moreover, it is consistent with the movement of the hinge region of the flaps, which is immediately adjacent. Unlike the expected movement of the flaps, this motion was not predicted in model building (Weber *et al.*, 1989; Lapatto *et al.*, 1989). Local asymmetry is also introduced as a consequence of different residues occupying the sites P3 and P3'. In the native enzyme Asp29 forms clear ionic interactions with Arg8 and Arg87. This interaction remains identical in one of the monomers, while in the second Arg8' is pushed away by Arg206 of the inhibitor which replaces it as the partner of Asp29' in the salt bridge. A bidentate ion pair between the carboxylate of Asp29' and the guanidinium group of Arg206 may explain why the Lys206 analog is a poor inhibitor (Toth and Marshall, unpublished). As a result of all these effects the two subunits forming the dimer, which were identical in the native structure since they were generated by the application of crystallographic symmetry, now deviate by 0.4 Å r.m.s. (comparison of $C\alpha$ positions).

The analysis of crystal packing indicated that the crystal form described here should be useful in the studies of the interactions of other inhibitors with the HIV-1 protease. The long axis of the protease molecule (vertical in Fig. 5) is aligned with the **x** axis of the unit cell, and large solvent channels are parallel to this axis. Both ends of the inhibitor face these channels and are not involved in crystal contacts, thus even inhibitors longer than the hexapeptide-based one described above may be accommodated.

Small co-crystals with several inhibitors were obtained under the conditions as described above. Data extending to 2.4Å were collected for the complex of HIV-1 PR with the heptapeptide:

Ac-Ser-Leu-Asn-Phe-[CHOH-CH_2N]-Pro-Ile-Val-OCH_3 (JG-365)

The scissile bond in this inhibitor is replaced by an unusual linkage containing an amino-alcohol group. The carbonyl functionality of the phenylalanine has been replaced with a hydroxyl group with resulting tetrahedral carbon geometry, and an additional methylene group was inserted between the Phe and Pro residues.

Crystals were isomorphous with crystals of the MVT-101 complex form (change in the unit cells parameters was smaller than 0.5%), so a difference Fourier map could be calculated based on the phases from the solved structure of this complex. Again, electron density for all residues of the heptapeptide was well defined. A model of the heptapeptide was fit to the initial density. The coordinates of JG-365 PR complex were refined using PROFFT. At the current stage of refinement the model includes 40 water molecules and the *R* factor is 0.208%.

A clear electron density indicating the position of the hydroxyl group can be seen (Fig. 7) between the two carboxyl groups of the active site aspartates, within hydrogen bonding distance. The hydrogen bond network between the inhibitor main chain oxygen and nitrogen atoms depicted in Figure 4 is the same for MVT-101 PR and JG-365 PR complexes for the six central residues; the nitrogen and carbonyl of P4 are exposed only to the solvent, while the hydroxyl of serine at P4 forms a hydrogen bond with Asp30. The importance of the interaction of the side chain at position P4 with the enzyme is consistant with an increase of K_i of the inhibitor from 0.5 nM to 21 nM upon removal of serine from the N-terminus of this heptapeptide (D. Rich *et al.*, 1989). Keeping in mind the preliminary nature of these data some general conclusions can be made:

1. The global, grossly symmetric change of the enzyme backbone upon inhibitor binding is the same for both peptides.
2. Hydrogen bonds between the main chain of the inhibitor and the enzyme involve subsites from P3 to P3' and are fully conserved in both complexes, supporting the view of their importance for the precise alignment of the ligand in the active site cleft (Sali *et al.*, 1989).
3. Side chain at position P4 is still important for binding.
4. The type of replacement for the scissile bond in the inhibitor may contribute significantly to the binding constant.

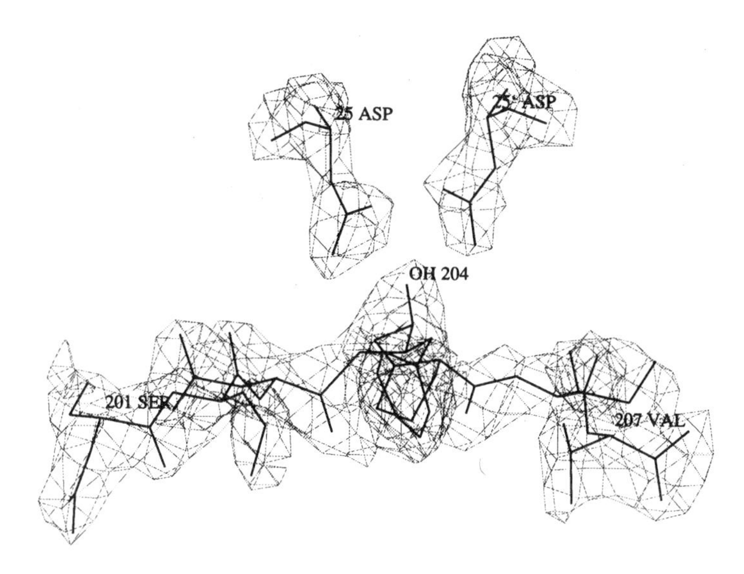

Figure 7.
The relative position of the heptapeptide JG-365 with respect to active site aspartates. The model is superimposed on a $|2F_O - F_C|\alpha_C$ electron density contoured at at 1σ level.

ACKNOWLEDGEMENTS

Research sponsored in part by the National Cancer Institute, DHHS, under contract NO. NO1-CO-74101 with BRI. The contents of this publication do not necessarily reflect the views or policies of the Department of Health and Human Services, nor does mention of trade names, commercial products, or organizations imply endorsement by the U.S. Government.

REFERENCES

Blundell, T.L., Jenkins, J., Pearl, L.H. Sewell, B.T. Pedersen, V. in 'Aspartic Proteinases and Their Inhibitors', (Kostka, V., ed.), Walter de Gruyter, Berlin, pp. 151-161, (1985).

Darke, P.L., Nutt, R.F., Brady, S.F., Garsky, V.M., Ciccarone, T.M., Leu, C.-T., Lumma, P.K., Freidinger, R.M., Veber, D.F. and Sigal, I.S. *Biochem. Biophys. Res. Comm.* **156**, 297-303 (1988).

Darke, P.L., Leu, C.-T., David, L.J., Heimbach, J.C., Diehl, R.E., Hill, W.S., Dixon, R.A.F. and Sigal, I.S. *J. Biol. Chem.* **264**, 23072312 (1989).

Erickson, J., Rao, J.K.M., Abad-Zapatero, C. and Wlodawer, A. unpublished.

James, M.N.G. and Sielecki, A.R. in 'Biological Macromolecules and Assemblies', (Jurnak, F.A. and McPherson, A. eds.), **3**, Wiley, New York, pp. 413-482 (1987)

Jaskólski, M., Miller, M., Rao, J.K.M., Leis, J. and Wlodawer, A. unpublished.

Kent, S.B.H. *Ann. Rev. Biochem.* **57**, 957–984, (1988).

Kohl, N.E., Emini,E.A., Schleif,W.A., Davis,L.J., Heimbach,J.C., Dixon,R.A.F., Scolnick,E.M. and Sigal,I.S. *Proc. Natl. Acad. Sci USA*, **85**, 4686–4690, (1988).

Lapatto, P., Blundell,T.L., Hemmings,A., Overington,J., Wilderspin,A., Wood,S., Merson,J.R., Whittle,P.J., Danley,D.E., Geoghegan,K.F., Hawrylik,S.J., Lee,S.E., Scheld,K.G. and Hobart,P.M. *Nature* **342**, 299–302, (1989).

Merrifield, R.B. *J. Am. Chem. Soc.* **85**, 2149–2154 (1963).

McKeever, B.M., Navia,M.N., Fitzgerald,P.M.D., Springer,J.P., Leu,C.-T., Heimbach,J.C., Herber,W.K., Sigal,I.S. and Darke,P.L. *J. Biol. Chem.* **264**, 1919–1921, (1989).

Miller, M., Jaskólski, M., Rao, J.K.M., Leis, J. and Wlodawer, A. *Nature* **337**, 576–579, (1989).

Miller, M., Schneider,J., Sathyanarayana,B.K., Toth,M.V., Marshall,G.R., Clawson,L., Selk,L., Kent,S.B.H. and Wlodawer,A. *Science*, **246**, 1149–1152, (1989).

Navia, M.A., Fitzgerald,P.M.D., McKeever,B.M., Leu,C.-T., Heimbach,J.C., Herber,W.K., Sigal,I.S., Darke,P.L. and Springer,J.P. *Nature*, **337**, 615–620, (1989).

Rich,D.M., Green,J., Toth,M.V., Marshall,G.R. and Kent,S.B.H., unpublished.

Sali, A., Veerapandian,B., Cooper,J.B., Foundling,S.I., Hoover,D.J. and Blundell,T.L. *EMBO J.*, **8**, 2179–2188, (1989).

Schneider, J. and Kent, S.B.H. *Cell*, **54**, 363–368, (1988).

Sielecki, A.R., Hayakawa,K., Fujinaga,M., Murphy,M.E.P., Fraser,M., Muir,A.K., Carilli,C.T., Lewicki,J.A., Baxter,J.D. and James,M.N.G. *Science*, **243**, 1346–1351, (1989).

Suguna, K., Padlan, E.A., Smith, C.W., Carlson, W.D., and Davies, D.R. *Proc. Natl. Acad. Sci. USA*, **84**, 7009–7013, (1987).

Szelke, M., Leckie,B., Hallett,A., Jones,D.M., Sueiros,J., Atrash,B., and Lever,A.F. *Nature*, **229**, 555–557, (1982).

Tam, J.P., Heath, W.F., Merrifield, R.B. *J. Am. Chem. Soc.*, **108**, 5242–5251, (1986).

Weber, I.T., Miller,M., Jaskólski,M., Leis,J., Skalka,A.M. and Wlodawer,A. *Science*, **243**, 928–931, (1989).

Wlodawer, A., Miller,M., Jaskólski,M., Sathyanarayana,B.K., Baldwin,E.T., Weber,I.T., Selk,L.M., Clawson,L., Schneider,J. and Kent,S.B.H. *Science*, **245**, 616–621, (1989).

11
Expression of HIV-1 *gag* and *pol* Gene Products using Recombinant Baculoviruses

Helen R. Mills, Hilary A. Overton and Ian M. Jones

Since its original description some years ago (Smith *et al.*,1983a, 1983b), recombinant baculoviruses have become an increasingly popular method for the expression of heterologous gene products. Genes of interest are introduced into the genome of the *Autographica californica* nuclear polyhedrosis virus (AcNPV) to give a helper independant recombinant virus that, upon infection of insect cells, produces the product encoded by the introduced gene in a regulated manner. Frequently, though not always, excellent yields of recombinant protein are produced (for a review see Luckow and Summers 1988a). Moreover, as the host expression background is a higher eucaryotic cell, many post-translational modificaitions such as myristylation, phosphorylation and glycosylation are efficiently carried out, leading to recombinant products that are often indistinguishable from their 'natural' counterparts. We have used this system to produce some ten different recombinant baculoviruses expressing various gene products encoded by both HIV-1 and HIV-2. Here we describe four of these recombinant viruses that produce products encoded by the gag and *pol* genes of HIV-1. We show that a feature of gag and *pol* translation in HIV-1 infected cells, the ribosomal frame shift, also occurs in insect cells leading to production of the HIV protease and specific cleavage of the *gag* precursor. The *gag* product, p55, and the protease can also be produced independently using recombinant baculoviruses and when co-infected into insect cells, cleavage of p55 *in trans* is observed. In the absence of cleavage by the protease, the p55 *gag* precusor molecule assembles into structures closely resembling the HIV-1 pre-core. When only p24 is expressed however, and despite very high expression levels, no such structures are observed suggesting that the primary amino acid sequences necessary for subunit interaction and assembly do not lie within the p24 coding region.

CONSTRUCTION OF RECOMBINANT BACULOVIRUSES

Recombinant baculoviruses were constructed using standard procedures (Summers and Smith, 1987). We have used two different transfer vectors for the generation of recombinant viruses, pAcYMI (Matsuura *et al.*, 1987) and pAcRP14 (Overton *et al.*, 1989). In the former vector, the complete transcription signals of the baculovirus polyhedrin promoter are maintained but the incoming insert must provide its own ATG initiation codon. In the latter plasmid, an ATG initiation condon (the ATG of the polyhedrin gene) is provided by the transfer vector itself allowing the expression of gene fragments that lack their own initiator ATG. It is important to note that, whilst use of the pAcYMI vector normally achieves high level expression of the cloned product (Matsuura *et al.*, 1987), sequences around the ATG of the inserted gene may modulate expression levels (Kozak, 1987; Luckow and Summers, 1988b). Use of pAcRP14 however bypasses this possibility as the ATG provided is preceeded by the upstream sequences that normally give highly efficient tra nslation of the polyhedrin mRNA. Thus, use of pAcRP14 should ensure high level protein production in all cases other than those in which transcription or translation of the insert coding region itself is intrinsically poor. Construction of *gag-pol*, *gag* and *prt* recombinant baculoviruses has been previously described (Overton *et al.*, 1989). Briefly, for the production of a *gag-pol* recombinant (Ac.*gag-pol*) a DNA fragment encoding the entire *gag* open reading frame and the first half of the *pol* open reading frame was cloned into pAcYMI and used to select polyhedrin - negative plaques. In this virus, translation of the *gag* open reading frame commences at the *gag* initiator ATG. In the construction of the *gag* and Protease viruses (Ac.*gag* and Ac.*prt* respectively) pAcRP14 was used as the transfer vector and the ATG it provided fused in frame to the relevant HIV open reading frame. For the *gag* recombinant virus, the ATG was fused to amino acid 15 of the p55 *gag* precursor protein leading to an expressed protein that lacks an authentic amino terminus. For the Protease virus the ATG was fused to the *pol* open reading frame at a position 64 amino acids prior to the protease coding region. However, as

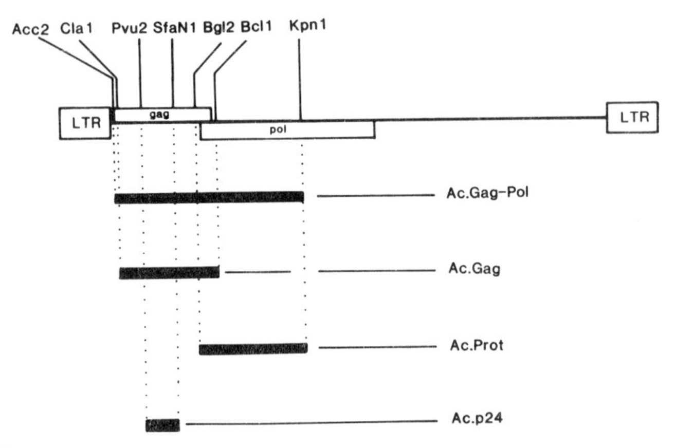

Figure 1.
Regions of the HIV - 1 genome used to produce recombinant baculoviruses. DNA fragments produced by the restriction enzymes indicated were cloned into transfer vectors pAcYM1 or pAcRP14 and used to generate polyhedrin negative plaques.

in the other expression systems expressed protease is capable of auto-excision (Debouck *et al.*, 1987, Hansen *et al.*, 1988, Le Grice *et al.*, 1988), we assume that the exact nature of the amino terminus of the expressed protease precursor is of little significance. To produce a recombinant baculovirus expressing p24 antigen (Ac.p24), a DNA fragment spanning the p24 coding region but also including part of p17 (12 amino acids) and p12 (12 amino acids) was fused in phase to the ATG of pAcRP14. Thus, in this virus, a p24 protein flanked by short regions of p17 and p12 is produced. The regions of the HIV genome used to produce each recombinant are summarized in figure 1.

GAG AND *POL* PRODUCT EXPRESSION BY RECOMBINANT BACULOVIRUSES

Once isolated, each recombinant virus was characterised for the presence of HIV antigens by Western blotting. *Spodoptera frugiperda* cells were infected with each recombinant virus at a multiplicity of infection of 10 and harvested at 2 or 3 days post infection when most recombinant

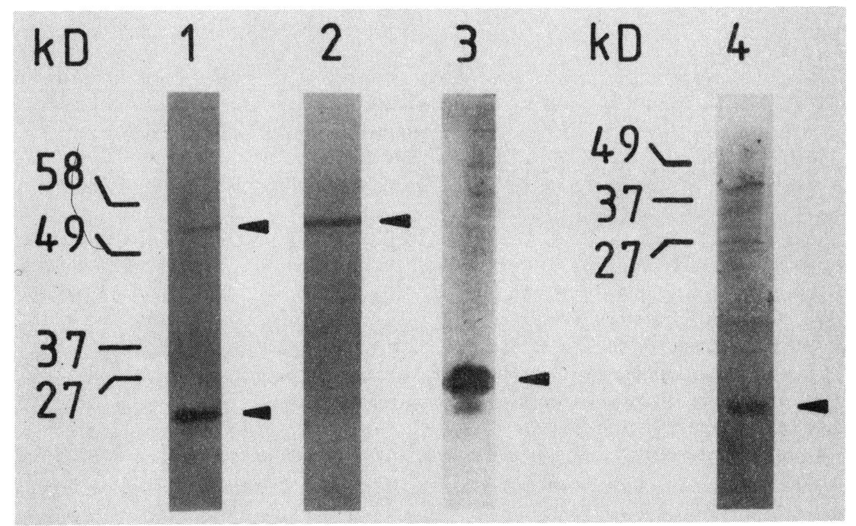

Figure 2.
Western blot analysis of the products produced by each recombinant virus. Panel 1 - cells infected with Ac.Gag-pol and products identified with an AIDS patient serum. Panel 2 - cells infected with Ac.*gag* and blotted with a patient serum. Panel 3 - cells infected with Ac.p24 and blotted with an anti-p24 monoclonal antibody. Panel 4 - cells infected with Ac.*prt* and blotted with a rabbit antiserum raised against a synthetic peptide spanning the protease active site. In all cases infected cells were harvested at two days post infection. Numbers on the left of panel 1 and panel 4 are molecular weights taken from a commercial pre-stained marker set (Sigma SDS-7B) and are in kilodaltons (kD). Note the change in scale for panel 4 compared to panels 1-3.

protein production is at a maximum (Summers and Smith 1987). Total cell lysates were fractionated on SDS-polyacrylamide gels and transfered to nitrocellulose membranes as described (Burnette, 1981). Membranes were incubated with HIV specific antisera and bound antibody detected by a second species alkaline phosphatase conjugate. The results are shown in Figure 2.

Each recombinant tested expressed HIV antigens with the mobility expected for the products encoded by the regions of the HIV genome incorporated. Ac.*gag-pol* produces a small fraction of p55 precursor whilst most of the antigen present is cleaved to p24 (panel 1). We have shown elsewhere that p17 is also present and that these antigens are correctly post-translationally modified (Overton *et al.*, 1989). The formation of the authentically cleaved *gag* proteins confirms that the ribosomal frame shift necessary to produce a functional HIV protease (Kramer *et al.*, 1986, Jacks *et al.*, 1988) occurs within the insect cell background (see also Madisen *et al.*,1987). Expression of the $p55^{gag}$ precursor alone by Ac.*gag* produces no cleaved products (panel 2) indicating that cleavage within Ac.*gag-pol* virus infected cells is HIV protease specific. Recombinant virus expressing p24 antigen (panel 3) produces an antigenically active protein of approximately 26 kD, slightly larger than authentic p24 (c.f. panel 1) as a result of the additional amino acids present at the amino and carboxy termini of this product. Finally, Ac.*prt* virus (panel 4) produces an antigenically active band at approximately 9kD corresponding to the mature, auto-excised HIV protease. Yields of the various *gag* and Pol products varied considerably. Total *gag* antigens in Ac.*gag-pol* virus infected cells were in the order of 1 mg per litre of 2×10^9 cells. Similar yields were obtained for the protease expressed by Ac.*prt*. Both the $p55^{gag}$ precursor and p24 were produced at substantially higher yield by their respective recombinant baculoviruses; p55 being produced at 10-50 mgs per litre of cells whilst p24 levels reached 100 mgs per litre or greater at three days post infection. Part of the reason for higher levels of expression by these latter two viruses is that each used the transfer vector pAcRP14 in which the signals necessary for efficient translation of messenger RNA are maintained (see above). However, the protease recombinant virus also employed pAcRP14 as transfer vector yet gave only low level expression. We assume from this result that accumulation of the protease to high levels is, in itself, detrimental to the insect cell.

ACTIVITY AND ASSEMBLY OF *GAG* AND *POL* PRODUCTS EXPRESSED IN INSECT CELLS

HIV protease activity within the insect cell was apparent by the observation of cleavage of the *gag* precursor within Ac.*gag-pol* recombinant baculovirus infected cells and by auto-excision of the protease from a larger *pol* open reading frame precursor in cells infected with Ac.*prt*. In addition, we have sought to demonstrate the activity of the protease *in trans* on the *gag* precursor when expressed in the same cell.

To do this insect cells were infected with either Ac.*gag* alone or with a 1:1 mixture of Ac.*gag* and Ac.*prt*. At two days post infection the cells were harvested and analysed for the presence of *gag* proteins by Western blotting. The results are shown in figure 3.

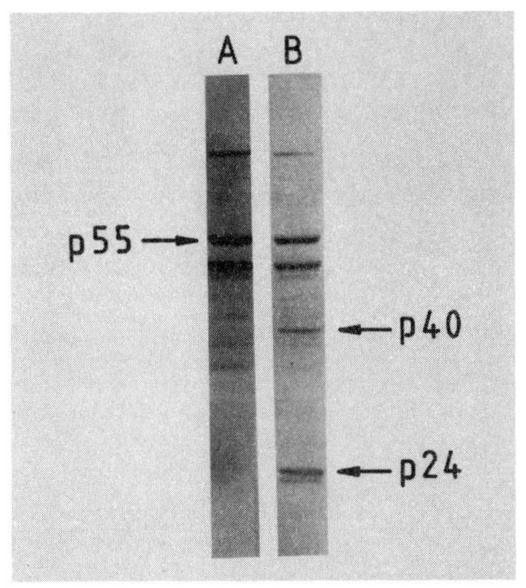

Figure 3.

Activity of the HIV protease *in trans*. *Spodoptera frugiperda* cells were infected with Ac.Gag alone (track A) or with a mixture of Ac.*gag* and Ac.*prt* (track B) at multiplicities of infection of 10 and harvested at two days post infection. Total cell extracts were analysed for the presence of Gag antigen by Western blotting using an anti-p24 monoclonal antibody. Specific cleavage products are indicated.

Cleavage of the p55^{gag} precursor is observed only in cells co-infected with the protease virus. Thus, co-infection of insect cells with two or more recombinant baculoviruses can deliver a number of recombinant proteins to the same cell and allow their interaction. Notwithstanding the demonstration of HIV protease activity *in trans*, Figure 3 also shows that large amounts of p55 precursor remain uncleaved during the co-infection. Moreover, in our early experiments, following detergent lysis of Ac.*gag* infected cells, we noted that essentially all the *gag* antigen segregated with the pellet following a clearing spin. These observations led us to investigate the physical state of the p55^{gag} precursor within the insect cell. To do this, insect cells infected with Ac. *gag* recombinant virus were harvested at two days post infection, fixed, sectioned and prepared for electron microscopy as described (Overton *et al.*, 1989). Typical results are shown in figure 4.

Ac.*gag* virus infected insect cells were found to contain large quantities of particulate material made up of bilayered spheres and hemispheres with approximate diameters of 90nm. Such structures could be

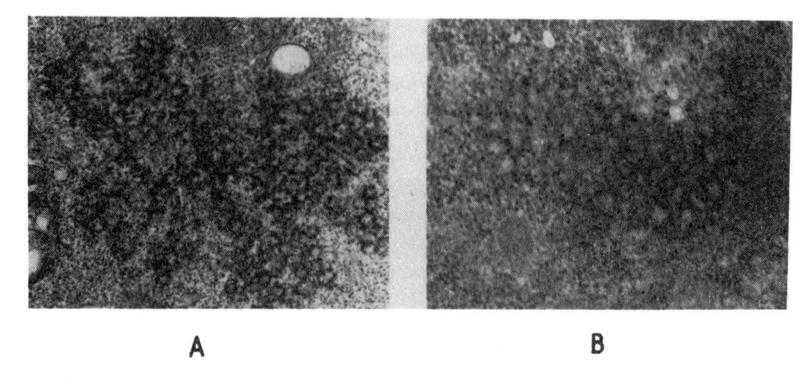

Figure 4.
Electron micrographs of thin sections through Ac.Gag infected cells. Panel A - section stained with uranyl acetate followed by lead citrate. Panel B - immunogold labelling with an anti-p24 monoclonal antibody followed by goat anti-mouse conjugated to 10nm gold particles.

specifically labelled with an anti-p24 monoclonal antibody (Figure 4 panel B) confirming their composition as *gag* antigen. The location of such structures within infected insect cells was predominantly cytoplasmic but occasionally some *gag* antigen was found in the nucleus. In appearance, these structures were similar to the HIV pre-core observed in HIV infected cells (Gelderblom *et al.*, 1988) and we conclude that the $p55^{gag}$ antigen, when expressed alone using recombinant baculoviruses, has the property of self assembly into retroviral core-like particles and that neither HIV RNA nor any other HIV protein is required for such assembly. We have also examined Ac.*gag-pol* recombinant baculovirus infected cells for the presence of core-like particles without success. However, as shown above, almost all the $p55^{gag}$ antigen in these cells is processed to p24 and p17 so the residual levels of uncleaved p55 may be too low to allow the direct observation of particles even if they form. Similarly, thin sections through insect cells producing only p24 failed to identify p55 - like particulate structures and, in keeping with this observation , we have found that the p24, unlike p55, is a soluble product following lysis by detergent or sonication (Figure 5). It is clear, therefore, that the primary amino acid sequences necessary for subunit assembly of the $p55^{gag}$ precursor do not lie wholly within the p24 region of the protein.

CONCLUSIONS AND FUTURE PROSPECTS

Recombinant baculoviruses have been used successfully to produce a number of the *gag* and *pol* proteins of HIV-1. A number of features of *gag-pol* gene expression in HIV-1 infected cells appear also to occur in insect cells infected with baculovirus recombinants. The ribosomal

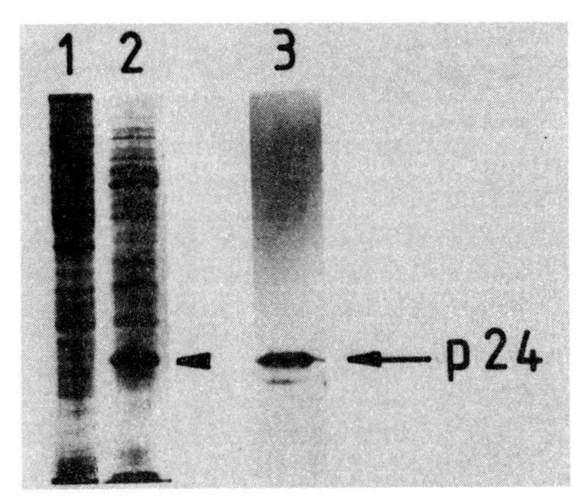

Figure 5.
Soluble p24 produced by Ac.p24. Infected cells were harvested at 2 days post infection, washed in isotonic saline and lysed in 0.1% Empigen BB (Albright and Wilson) in isotonic saline for 20mins at 0°C. Lysates were clarified by centrifugation at 100,000g for 1hr and the supernatants analysed by SDS-PAGE followed by coomassie staining or Western blotting. Track 1 - uninfected cell lysate. Track 2 - Ac.p24 infected cell lysate. Track 3 - Western blot of track 2 with anti p24 monoclonal antibody. The position of p24 antigen is indicated.

frame shift, auto-excision of the HIV protease and assembly of the p55 *gag* precursor have all been observed in infected insect cells. The assembly of the p55 precursor in particular is an observation that suggests the baculovirus expression system as a useful method for the study of the process of HIV pre-core assembly. Since our original observation of *gag* p55 assembly (Overton *et al.*, 1989), two further reports have used recombinant baculoviruses to study the self assembly of *gag* antigen; p55 of HIV (Gheysen *et al.*, 1989) and p57 of SIV (Delchambre *et al.*, 1989). In both of these elegant studies, it was shown that the complete *gag* antigen when expressed in insect cells was myristylated at its amino terminus, assembled into pre-core structures and secreted into the surrounding media. Loss of myristylation resulted in cell bound particles identical to those shown in figure 4. Our own *gag* virus described earlier produces a *gag* protein that has lost amino - terminal amino acids including the signal for myristylation and, accordingly, particles produced by this recombinant virus are not secreted. Moreover, we can conclude that the amino terminal 14 amino acids of p55 are not required for partic le assembly. Gheysen *et al.* (1989) also observed that truncation of the p55 molecule at the carboxy terminal end of p24 led to membrane accumulation but not to pre-core assembly. We have shown here that p24 antigen, when expressed alone, does not give rise to aggregated structures but is soluble. Thus, it appears that the primary sequences necessary for p55 subunit assembly lie outside of the p24 coding region possibly in the region coding for p12. Use of the baculovirus system to express a comprehensive set of *gag* truncations might allow the sites of subunit interaction to be defined. Molecular mimics of such regions could then be considered as agents for therapeutic intervention aimed at pre-core disruption rather like those agents already described that block enzyme subunit assembly (Dutia *et al.*, 1986, Cohen *et al.*, 1986). We have also expressed p55^{gag} antigen with the HIV protease in the Ac.*gag-pol* recombinant virus. However, in this case almost all the p55^{gag} antigen is processed to the mature *gag* antigens p24, p17 and p12 and, as a consequence, no particle formation is observed. Similar results were also obtained by Gheysen *et al.* (1989). The fact that protease activity is apparently uncontrolled in these cells might suggest that there are mechanisms within HIV-1 infected cells that limit protease activity prior to virion assembly when maturation of the *gag* precursor is thought to occur (Katsumoto *et al.*, 1987, Gelderblom *et al.*, 1988). In addition to *gag* subunit assembly, the HIV protease must itself assemble to form a dimer in order to become active (Navia *et al.*,1989). In HIV infected cells, dimerisation must first occur with *gag-pol* fusion proteins as these are the primary sources of HIV protease. It is possible that *gag-pol* subunit interaction occurs via the *gag* domains in a similar way to p55^{gag} assembly. However, we have noted that, when expressed from only the *pol* coding region using Ac.*prt*, the protease is active suggesting it has intrinsic dimerisation properties. Thus, interaction between *gag-pol* proteins necessary for protease activation may be quite different from the interaction of *gag* proteins alone. When *Spodoptera frugiperda* cells are co-infected with both Ac.*gag* and Ac.*prt*, cleavage of the *gag* precursor *in trans* was observed to take place (cf. Figure 3) and we have not yet examined such co-infected cells for the physical state of the p55^{gag} antigen. However, it is possible that ultrastructural studies of such co-infected cells might allow the study of the intermediate stages of core maturation as the pre-core is digested within the insect cell by the protease provided *in trans*.

ACKNOWLEDGMENTS

Our work was funded by the UK Medical Research Councils Aids directed program.

REFERENCES

Adams, S.E., Dawson, K.M., Gull, K., Kingsman, S.M. and Kingsman, A.J. *Nature*, **329**, 68–70 (1987).

Burnette, W.N. *Anal. Biochem.*, **112**, 195–203 (1981).

Cohen, E.A., Gandeau, P., Brazeau, P. and Langelier, Y. *Nature*, **321**, 441–443 (1986).

Debouck, C., Gorniak, J.G., Strickler, J.E., Meek, T.D., Metcalf, B.W. and Rosenberg, M. *Proc. Natl. Acad. Sci. USA*, **74**, 8903–8906 (1987).

Delchambre, M., Gheysen, D., Thines, D., Thiniart, C., Jacobs, E., Verdin, E., Horth, M., Burny, A. and Bex, F. *EMBO J.*, **8**, 2653–2660 (1989).

Dutia, B.M., Frame, M.C., Subak-Sharpe, J.H., Clark, W.N. and Marsden, H.S. *Nature*, **321**, 439–441 (1986).

Gelderblom, H.R., Ozel, M., Hausmann, E.H.S., Winkel, T., Pauli, G. and Koch, M.A. *Micron. Microsc.*, **19**, 41–60 (1988).

Gheysen, D., Jacobs, E., de Foresta, F., Thiriart, C., Francotte, M., Thines, D. and de Wilde, M.D. *Cell*, **59**, 103–112 (1989).

Hansen, J., Bilich, S., Schulze, T., Sukrowm S. and Moelling, K. *EMBO J.*, **7**, 1785–1791 (1988).

Jacks, T., Power, M.D., Masianz, F.R., Luciw, P.A., Barr, P.J. and Varmus, H.E. *Nature*, **331**, 280–283 (1988).

Katsumoto, T., Hattori, N. and Karimura, T. *Intervirology*, **27**, 148–153 (1987).

Kozak, M. *J. Mol. Biol.*, **196**, 947–950 (1987).

Kramer, R.A., Schaber, M.D., Skalka, A.M., Gangluy, K., Wong-Staal, F. and Reddy, E.P. *Science*, **231**, 1580–1584 (1986).

Le Grice, S.F.J., Mills, J. and Mous, J. *EMBO J.*, **7**, 2547–2553 (1988).

Luckow, V.A. and Summers, M.D. *Biotechnology*, **6**, 47–55 (1988a).

Luckow, V.A. and Summers, M.D. *Virology*, **167**, 56–71 (1988b).

Madisen, L., Travis, B., Hu, S-L. and Purchio, A.F. *Virology*, **158**, 248–250 (1987).

Matsuura, Y., Possee, R.D., Overton, H.A. and Bishop, D.H.L. *J. Gen. Virol.*, **68**, 1233–1250 (1987).

Navia,M.A., Fitzgerald,P.M.D., McKeever,B.M., Leu,C-T., Heimbach,J.C., Herber,W.K., Sigal,I.S., Darke,P.L. and Springer,J.P. *Nature*, **337**, 615 – 620 (1989)

Overton, H.A., Fujii, Y., Price, I.R. and Jones, I.M. *Virology*, **170**, 107–116 (1989).

Smith, G.E., Fraser, M.J. and Summers, M.D. *J. Virol.*, **46**, 584–593 (1983a).

Smith, G.E., Summers, M.D. and Fraser, M.J. *Mol. Cell. Biol.*, **3**, 2156–2165 (1983b)

Summers, M.D. and Smith, G.E. 'A Manual of Methods for Baculovirus vectors and Insect Cell Culture Proceedures', Texas Agricultural Experimental Station Bulletin No. 1555 (1987).

12
Morphogenesis of Retroviruses: A Complex Role for the Matrix (MA) Protein in Assembly, Stability and Intracellular Transport of Capsid Proteins

Sung Rhee and Eric Hunter

Members of the retrovirus family, are spherical, RNA-containing viruses thatare enveloped by a membrane acquired during budding from the plasma membrane of an infected cell. They have common morphological features that reflect similar mechanisms of morphogenesis. The viral envelope is studded with glycoprotein knobbed-spikes, and in nascent particles surrounds an icosahedral capsid assembled mostly from internal structural proteins encoded by the viral *gag* gene. In the mature virion, this polyhedral capsid consists of an envelope-associated outer shell that in turn encloses an inner ribonuclear-protein core composed of genomic RNA and structural protein subunits (Nermut *et al.*, 1972; Marx *et al.*, 1988). The morphology of the core is somewhat virus specific and may be spherical (Rous sarcoma virus, RSV), rod-shaped (Mason-Pfizer monkey virus, M-PMV) or cone-shaped (human immunodeficiency virus, HIV) in electron micrographs.

Since final assembly of a retrovirus occurs at the plasma membrane of infected cells, viral proteins must be transported, using the host cell's own transport machinery, to the plasma membrane from the subcellular compartments where they are synthesized and modified. Envelope glycoproteins are synthesized as membrane-spanning proteins on polysomes associated with the endoplasmic reticulum and are then transported through the secretory pathway of the cell to the plasma membrane. The general details of the secretory pathway have been established in studies of secreted cellular proteins (Palade, 1975) and it is likely that retroviral glycoproteins utilize the same basic mechanisms and signals for transport to the cell surface (Hunter, 1988). The internal capsid proteins, on the other hand, are synthesized as cytoplasmic proteins on free polysomes and are then transported, by yet to be defined pathways, to the inner face of the plasma membrane where virus budding occurs.

Two quite distinct pathways appear to operate for intracellular transport of the viral capsid proteins, which result in two different morphogenic processes for different retrovirus types. In C-type retroviruses such as murine leukemia virus and RSV the capsid is assembled, from individually transported capsid precursor polyproteins, as the virus buds from the plasma membrane. In contrast in B- and D-type retroviruses, exemplified by mouse mammary tumor virus and M-PMV, an immature capsid is preassembled within the cytoplasm prior to transport to the cell membrane. In the latter viruses, therefore, only the preassembled capsid appears to have the necessary signals for intracytoplasmic transport, whereas these presumably exist on individual capsid proteins of C-type viruses. In either case little is known about the details and mechanisms of the transport process. However in both C-type (Kawai and Hanafusa, 1973; Linial *et al.*, 1980) and D-type viruses (Rhee and Hunter, manuscript in preparation), viral glycoproteins have been shown to play no required role in the transport of either capsid proteins or preassembled capsids to the membrane. This supports the concept that intrinsic signals exist in capsid proteins for their intracytoplasmic transport.

Gag proteins are synthesized and assembled into an immature capsid as precursor polyproteins, which are then proteolytically cleaved to the structural proteins of the mature capsid during or shortly after virus budding (Figure 1).The function of each of these mature polypeptides in the various stages of retrovirus replication is not fully understood. Since in B- and D-type viruses the processes of assembly and intracytoplasmic transport of capsid are temporally and spatially unlinked, these viruses provide an excellent system for identifying the functional roles of *gag* proteins in virus replication. Over the past several years we have developed the M-PMV system by characterizing viral protein biosynthesis and by molecularly cloning and sequencing an infectious virus genome (Barker *et al.*, 1985; Barker *et al.*, 1986; Sonigo *et al.*,1986). With this molecularly characterized and amenable system, we are currently studying the functional roles of viral *gag*-gene encoded proteins in virus assembly and maturation.

Studies with several retroviruses have suggested that the amino-terminal protein of the *gag* gene product associates with the virus envelope (Cardiff *et al.*, 1978; Marcus *et al.*, 1978; Pepinsky and Vogt, 1979). Therefore thisprotein has been designated as the retroviral matrix (MA) protein by analogy with the lipid-associated M-protein of vesicular stomatitis virus and other negative-stranded viruses (Pepinsky and Vogt, 1979). The MA protein of a retrovirus presumably functions in mediating interactions between the other *gag* proteins and the cell membrane. After the cleavage from the precursor during maturation it would thus form the envelope-associated outer shell of the polyhedral capsid and surround the inner core. Indeed immuno-electron microscopic studies by Gelderblom *et al.* (Gelderblom *et al.*, 1987) showed that the matrix protein of HIV, p17, is a component of the envelope-associated icosadeltahedral capsid that lines the inner surface of virus envelope. Because of its association with the membrane, the MA protein has also been postulated to interact with the virus envelope glycoproteins at the plasma membrane during budding (Cardiff *et al.*, 1978; Montelaro *et al.*, 1978). This concept is supported by chemical cross-linking studies; with Rous sarcoma virus, Gebhardt *et al.* (Gebhardt *et al.*, 1984) showed that the matrix protein, p19,could be crosslinked to the transmembrane glycoprotein, gp37. However, the nature of the interaction between these two proteins has not been defined and it is not clear whether such interactions are essential for glycoprotein incorporation into virions.

The *gag* polyprotein, Pr78^{gag}, encoded by the M-PMV *gag* gene, is proteolytically cleaved by the viral protease to yield the six capsid proteins of mature virion (p10, pp16-18, p12, p27, p14, and p4) (Bradac and Hunter, 1984; Henderson *et al.*, 1985). Because it is in an analogous position to the matrix proteins of other retroviruses (Figure 1), p10 has been designated the matrix protein of M-PMV (Bradac and Hunter, 1986). We describe here moleculargenetic studies on the role of p10(MA) in M-PMV capsid assembly, intracellular transport, and maturation.

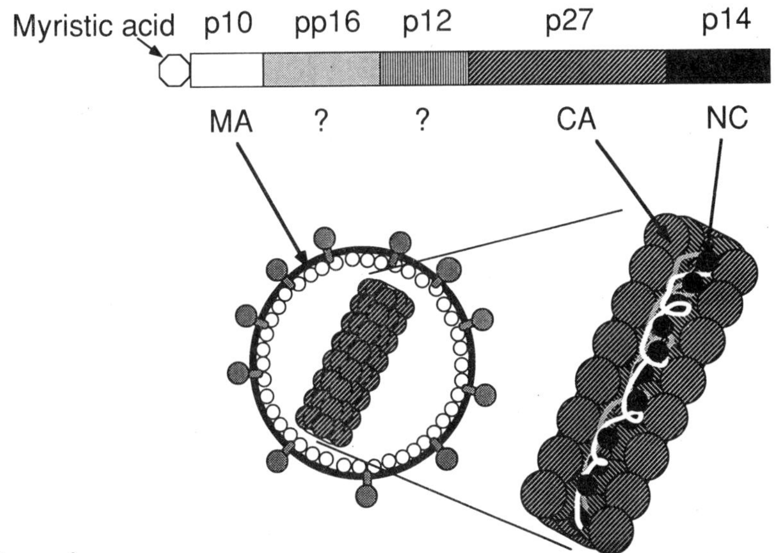

Figure 1.
Schematic diagram of the arrangement of the structural proteins within the Gag precursor polyprotein and in the virion.

RESULTS AND DISCUSSION

To obtain a better understanding of the functional roles of p10 in the assembly and maturation of M-PMV, three different mutational approaches have been followed within the p10 coding region:

1) The construction of point mutants by oligonucleotide-directed mutagenesis to create myristylation defective mutants of p10.

2) The construction of specific, in-frame deletion mutations within the p10 coding region using oligonucleotide-directed mutagenesis.

3) The introduction of random point mutations within the MA coding domain using sodium bisulphite modification of cytosines in this region.

In each case, *in vitro* mutagenesis was carried out on a fragment of the *gag* gene cloned into a single stranded bacteriophage. Verified mutant genes were then transferred back into an infectious proviral DNA clone which could be introduced into both HeLa and COS-1 cells for biological and molecular phenotyping studies. The positions of changed amino acids within p10 and the phenotypes of each mutant are summarized in Table 1.

Table 1.
Summary of the phenotypes of matrix protein mutants of M-PMV.

Mutants	Mutations	Precursors Stability	Capsid Assembly	RT in Media	Infectivity Cell-Free
Myristylation-defective Mutants					
A-1p10	G(2) → V	+++	+++	–	–
A-2p10	G(2) → E	+++	+++	–	–
A-3p10	G(2) → A	+++	+++	–	–
Deletion Mutants					
D-11/94-p10	ΔY(11)-E(94)	–	+/–	–	–
D-12/32-p10	ΔV(12)-L(32)	–	+/–	–	–
D-33/53-p10	ΔK(33)-I(53)	–	+/–	–	–
D-54/74-p10	ΔK(54)-K(74)	–	+/–	–	–
D-79/85-p10	ΔA(79)-L(85)	–	+/–	–	–
D-8/58-p12	ΔE(8)-W(58)	+++	–	–	–
Sodium Bisulphite Mutants					
Wild-Type like Mutants					
T21I-p10	T(21) → I	+++	+++	+++	+++
T41I-p10	T(41) → I	+++	+++	+++	+++
T211-p10	T(21) → I	+++	+++	+++	+++
Stability Mutants					
P43L-p10	P(43) → L	–	+/–	–	–
P72S-p10	P(72) → S	+	+	–	–
p43L/S81F-p10	P(43) → L S(81) → F	–	+/–	–	–
P43S/T69I /A79V-p10	P(43) → S T(69) → I A(79) → V	+	+	–	–
Transport Mutants					
A18V-p10	A(18) → V	+++	+++	–	–
A79V-p10	A(79) → V	+++	+++	–	–
T69I-p10	T(69) → I	+++	+++	++	++
Assembly Mutants					
T41I/T78I-p10	T(41) → I T(78) → I	+++	++	+	+
T21I/R55W-p10	T(21) → I R(55) → W	+++	*–	++	–

Myristylation is Required for Intracytoplasmic Transport but not for Assembly of D-type Retrovirus Capsids.

As with other mammalian retroviruses, the *gag* polyprotein and the matrix protein, p10, in mature virions of M-PMV are myristylated (Schultz and Oroszlan, 1983). Myristic acid is added to an amino-terminal glycine residue of the *gag* polyprotein immediately after cotranslational removal of the terminal methionine. On the basis of the proposed membrane binding function of the matrix protein, myristic acid might be expected to be primarily responsible for the nonpolar interaction between the cell membrane and the virus capsid. However whether myristylation of the *gag* polyprotein might also be essential for assembly and transport of the capsid was not known.

In order to investigate the role of myristylation in this stage of D-type retrovirus morphogenesis, a point mutation, which alters the codon for the normally myristylated glycine to one for valine, was introduced into the M-PMV DNA proviral genome through the use of oligonucleotide-directed mutagenesis. In cells expressing the (A-1p10) mutant genome, non myristylated *gag* polyproteins were synthesized at normal levels and assembled into capsids (Table 1). Thus myristic acid addition to the *gag* polyprotein is not required forthe intracytoplasmic assembly of virus capsids. However there was nosubsequent proteolytic processing of the *gag* polyproteins or budding of virus particles into the culture medium. Instead large accumulations of nonmyristylated mutant capsids were detected in Western blots and thin-section electron micrographs of infected cells (Figure 2B). The majority of these capsids were found deep in the cytoplasm and none were observed near the plasma membrane. This result strongly suggested that myristylation provides a critical signal for intracytoplasmic transport of the D-type virus capsid to the site of budding and release; the absence of such an intracytoplasmic transport signal would explain the prominent accumulation of capsids in the cytoplasm of mutant-genome expressing cells (Rhee and Hunter,1987). However these experiments did not differentiate between a direct signalling role for myristic acid residues and a more subtle activation of cryptic signals within the precursor molecule. In an attempt to address thesepossibilities additional mutations were engineered into the the MA coding domain.

The Native Structure of p10 is Important for *Gag* Polyprotein Stability and Capsid Assembly.

Five different M-PMV mutants with in frame deletions within the p10 codingregion were made by using oligonucleotide-directed mutagenesis (Table 1). The large deletion mutant, D-11/94-p10, in which almost the entire coding region of p10 has been deleted, retains the amino-terminal ten amino acids and the carboxyl-terminal six amino acids of p10 in order to leave the myristic acid addition site and the proteolytic processing site between p10 and pp24 intact. The remaining deletions result in internal truncations of p10 of from 7 to 21 amino acids.

In cells transfected with these deletion-mutant genomes, mutant *gag* polyproteins were synthesized and modified with myristic acid in a normal manner. However these polyproteins were so unstable that half of the newly synthesized polyproteins disappeared within 1 hr without being processed into mature capsid proteins, in contrast to wild-type polyproteins which have aprocessing half-life of 2.5-3.0 hr. These deletion mutations in p10 thus appear to result in a structural alteration of the *gag* polyproteins which modifies their intracytoplasmic stability. Furthermore, these mutant *gag* polyproteins were assembled with very low efficiency into capsids in the cytoplasm of the mutant-infected cells, and since they were neither released from, nor accumulated in the cells, those mutant capsids which do assemble are presumably also unstable and degraded. The fact that all of the deletion mutants had similar phenotypes is most easily explained by postulating a critical structural role for p10 in both stabilizing and facilitating assembly of the immature intracytoplasmic capsids. Loss of stability might result from the mutant polyprotein precursor being unable to attain a favorable conformation and/or the introduction of an unfavorable conformation which could not be accommodated in the folded structure. Consequently this could interfere with the specific intermolecular interactions necessary for capsid assembly. In contrast to these observations with p10, deletion of 2/3 of the p12 coding domain (D-8/58-p12, Table 1) has no effect on precursor protein stability even though such a mutation blocks capsid assembly (unpublished data). Thus the matrix protein of M-PMV appears to play a crucial role incorrectly folding *gag* polyproteins, which in turn stabilizes the molecules and allows recognition for self-assembly into capsids.

Point Mutations within the Matrix Protein Affect the Process of Assembly and Intracytoplasmic Transport of Capsids.

The above studies on M-PMV morphogenesis showed that modification of the *gag* precursor protein by addition of myristic acid is required for capsid transport to the plasma membrane and that a native MA protein is required to fold the *gag* polyproteins into a conformation that confers stability and allows the intermolecular interactions involved in self-assembly to occur. To define further the functions of the matrix protein in the virus replication, we have carried out sodium bisulfite mutagenesis on the MA coding region to randomly substitute amino acids within the p10 coding region of M-PMV. Four classes of mutant phenotypes have been observed in these studies:

a). Mutants with a wild-type phenotype.

Mutants with single amino acid substitutions at positions 21 (threonine to isoleucine), 41 (threonine to isoleucine), and 57 (arginine to tryptophan) within the matrix protein, and which have been named respectively T21I-p10, T41I-p10, and R57W-p10 (Table 1) showed normal levels of protein biosynthesis and replication. No differences in cell-free infectivity from wild type virus were observed. Thus these amino acid substitutions have no effect on D-type virus assembly and maturation. They presumably do not critically participate in the signals proposed for intracytoplasmic retention of the *gag* polyprotein, assembly of capsids, and transport to the membrane.

b). Mutants which synthesize unstable *Gag* polyproteins analogous to those of p10 in-frame deletion mutants.

There are five proline residues within the p10 coding region of M-PMV. In two of the bisulfite mutants, proline residues at amino acid positions 43 and 72 were independently substituted to leucine (P43L-p10) and serine (P72S-p10), respectively (Table 1).

Mutant P43L-p10 transfected cells synthesized normal levels of the *gag* polyproteins but they are very unstable so that the majority of them are degraded prior to capsid assembly. This phenotype is identical to that of the in-frame deletion mutants described above, indicating that the proline residueat position 43 is a critical amino acid for a native, stable conformation of the M-PMV *gag* polyprotein. Mutant P43L/S81F-p10 which has two amino acid substitutions at positions 43 (proline to leucine) and 81 (serine to phenylalanine) showed the same highly unstable phenotype.

While mutant P72S-p10 also expressed an unstable *gag* polyprotein, this molecule was more stable than those with a leucine substitution. The intermediate stability of the protein may result from the introduction of serine residue instead of leucine at the proline position, since a triple mutant P43S/T69I/A79V-p10 with substitutions at positions 43 (proline to serine), 69 (threonine to isoleucine) and 79 (alanine to valine) showed a similar intermediate in stability (Table 1). In addition the mutant *gag* polyproteins of P72S-p10 and P43S/T69I/A79V-p10 assembled capsids more efficiently than P43L-p10 but they were neither released into the medium nor accumulated in the cell cytoplasm. This suggests that even though mutant capsids are formed, the modified ternary structure of such capsids prevents intracytoplasmic transport and subjects them to protein degradation.

The results from these point mutants support the concept that the matrixprotein of M-PMV plays an important role in allowing the *gag* polyprotein to fold in a stable, transportable conformation. It is possible that 'proline turns' within the MA region of the polyprotein play a critical role in the folding process since each of the unstable deletion mutants described above removes a proline residue. These early stages of D-type retrovirus morphogenesis thus appear to be very sensitive to changes in the tertiary structure of *gag* polyproteins and contrast that of the C-type viruses where insertion and deletion mutations in the MA coding region have little effect on budding and particle release from the cell (Voynow and Coffin, 1985; J. Wills, personal communication).

c). Mutants which are defective in intracytoplasmic transport of assembled capsids.

Two mutants, A18V-p10 and A79V-p10 which both have a single amino acid change of alanine to valine (Table 1) exhibited a phenotype similar to that of the myristylation defective mutant (A-1p10) described above. Cells expressing either mutant genome assembled myristylated polyprotein precursors into capsids in the cytoplasm but did not release viruses into the culture medium. Instead the immature capsids accumulated in the cytoplasm and none were found associated with the plasma membrane.

Since the A18V-p10 and A79V-p10 mutant capsids are assembled frommyristylated precursor molecules, it is clear that myristylation itself is not sufficient to function as a signal for directing the intracytoplasmic transport of self-assembled capsids to the plasma membrane. Indeed it seems possible that myristic acid addition may result in a necessary conformational change that renders the capsid transport-competent; mutations at amino acid residues 18 and 79 might be postulated to prevent the acquisition of such competency either by preventing this conformational change or by altering the signal itself .

In mutant T69I-p10 a threonine residue at position 69 was changed to an isoleucine. This mutant showed normal levels of *gag* polyprotein biosynthesis and efficient capsid assembly, but processing of mutant *gag* polyproteins was much slower than that of wild-type polyproteins and pulse-labeled *gag* proteins were only detected in the culture fluid after a significant delay. Since processing of retroviral *gag* polyproteins only occurs during or after virus release from the cell, it is possible that the phenotype of mutant T69I-p10 results from slow intracytoplasmic transport of capsids to the plasma membrane. This possibility was supported by electron microscopic analyses ofmutant transfected cells; significant numbers of mutant capsids were seen dispersed through the cytoplasm but accumulations of particles similar to those seen in A-1p10 or A18V-p10 and A79V-p10 were not observed . In this case then, it is possible that the point mutation either delays the formation of transport competency or interferes directly with capsid association with the transport machinery.

d). Mutants which show an altered process of capsid assembly.

Two mutants were obtained that showed altered processes in capsid assembly (Table 1). The first of these, mutant T41I/T78I-p10, assembles capsids in the cytoplasm but the process of assembly takes longer than in the wild-type virus. In pulse-labeled, wild-type transfected cells, *gag* precursor molecules are initially found in a soluble, non-particulate form in the cytosol. With a half-life of approximately 45 min. these radiolabeled molecules are incorporated into pelletable capsids. In contrast in cells expressing theT41I/T78I-p10 mutant genome, formation of particulate precursors is significantly delayed and less efficient. Thus these cells synthesize normal levels of stable *gag* polyproteins but release less virus into the culture medium. This mutation then either interferes with the transport of precursors to the site of capsid assembly or reduces the co-operative incorporation of *gag* polyproteins into the nascent capsid itself.

The second mutant T21I/R55W-p10 has two amino acids substitutions at positions 21 (threonine to isoleucine) and 55 (arginine to tryptophan) that confer an unexpected phenotype on this variant virus. Cells transfected with this mutant genome synthesized normal levels of *gag* polyproteins which were processed somewhat more rapidly than in wild-type infected cells, and released substantial amounts of non-infectious virions into the culture medium. However in contrast to those of wild type infected cells precursor proteins could not be shown to assemble into intracytoplasmic capsids by Western blot analyses. This was confirmed by monitoring the assembly of intracytoplasmic capsids, as described earlier, by the incorporation of pulse-labeled *gag* polyproteins into the pelletable fraction of 0.5% Triton X-100 solubilized cells. In wild-type infected cells about 25% of labeled *gag* polyproteins were incorporated into capsids during the pulse labeling period and increasing amounts of them were incorporated during a chase. In contrast no T21I/R55W-p10 mutant *gag* polyproteins were detected in the pelletable fraction even after a 4.5 hr chase. These results suggested that this mutant D-type retrovirus did not form intracytoplasmic capsids but instead assembled virions at the plasma membrane and electron microscopic studies of mutant transfected cells conclusively showed that a majority of the mutant *gag* polyproteins were assembled into capsids at the plasma membrane through a morphogenic process similar to that of C-type retroviruses (Figure 2D). Thus the two amino acid substitutions in the matrix protein in mutantT21I/R55W-p10 appear to convert its morphogenesis from that of a D-typeretrovirus to one of a C-type retrovirus. This exciting and unexpected finding suggests that the differences in the site of capsid formation of B- and D-type retroviruses versus C-type retroviruses can be modulated by transport signals in the newly synthesized *gag* polyproteins. In D-type retroviruses such signals might act to arrest the proteins in the cytoplasm so that they can self-assemble into immature capsids. Since mutant T21I/R55W-p10 efficiently assembles its capsids at the plasma membrane concurrently with the budding process, it is clear that the mutant *gag* polyproteins are not assembly defective. It seems more likely that they are inefficiently arrested within the cytoplasm and that the *gag* polyproteins are transported as individual molecules to the plasma membrane where they can associate with membrane and self-assemble into capsids

CONCLUSIONS

In summary, we have utilized a molecular genetic approach to investigate the role of the MA protein of M-PMV in virus replication. These mutational analyses show that modifications to this myristylated, membrane-associated protein can have multiple effects on the assembly, transport and release of capsids from M-PMV infected cells.

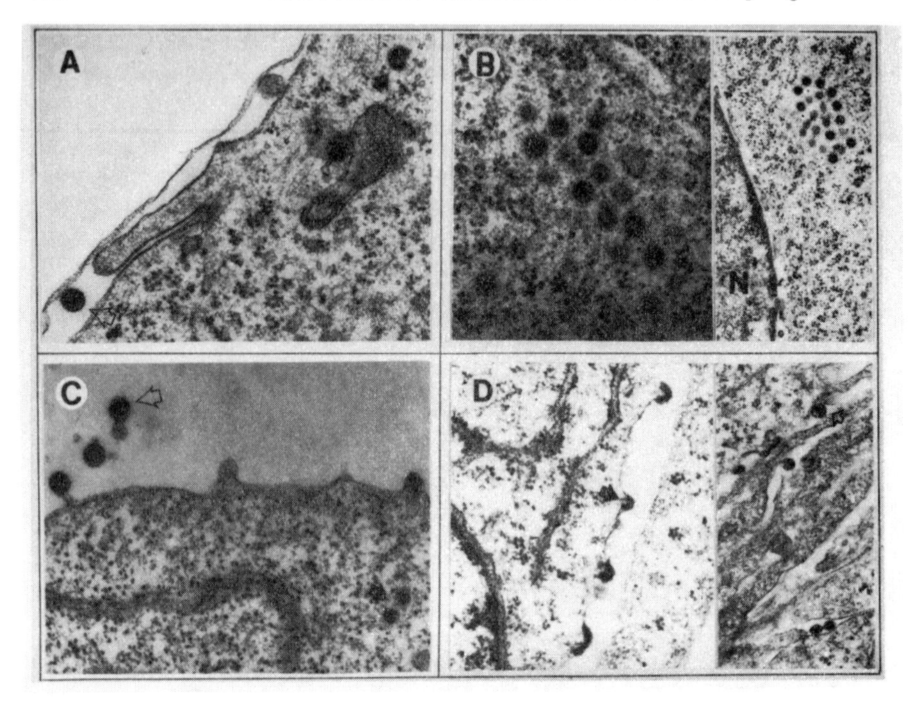

Figure 2.
Electron micrographs of wild-type and mutant genome transfected cells. (A, B) The thin sections of HeLa cells transfected with wild-type (A) and myristylation defective mutant (A-1p10) genomes: (A) an isolated assembling capsid (solid arrow) and a released mature virion (open arrow) are seen in wild-type infected cells. (B) In contrast, large accumulations of capsids can be seen deep in the cytoplasm of mutant genome expressing cells. No released or budding particles were observed in any section of mutant genome expressing cells. (C, D) The thin sections of COS-1 cells transfected with wild-type (C)and mutant T21I/R55W-p10 (D) genomes: (C) preassembled intracytoplasmic capsids (solid arrow) are seen in the cytoplasm of wild-type infected cells. Both released mature virions (open arrow) and particles budding from the membrane can be seen. (D) assembling capsids (solid arrow) can be seen at the plasma membrane of mutant T21I/R55W-p10 expressing cells. Mature virions (open arrow) are released from mutant-genome transfected cells.

Although our initial experiments showed a critical role for myristylation in signalling transport of assembled capsids from their site of assembly to the site of budding from the plasma membrane, the identification of single point mutations within p10 that confer the same phenotype argues against myristic acid being the primary transport signal. Indeed in the picornaviruses and papovaviruses where capsid proteins are also modified by myristic acid, this hydrophobic residue has been located within the interior of the capsid (Hogle *et al.*, 1985; Paul *et al.*, 1987; Rossman *et al.*, 1985; Schmidt *et al.*, 1989). It is possible then that the primary purpose of the fatty acid addition is to induce or stabilize a specific conformation within the capsid that confers a state of transport competence, which in the absence of myristylation or in the presence of certain mutations cannot occur.

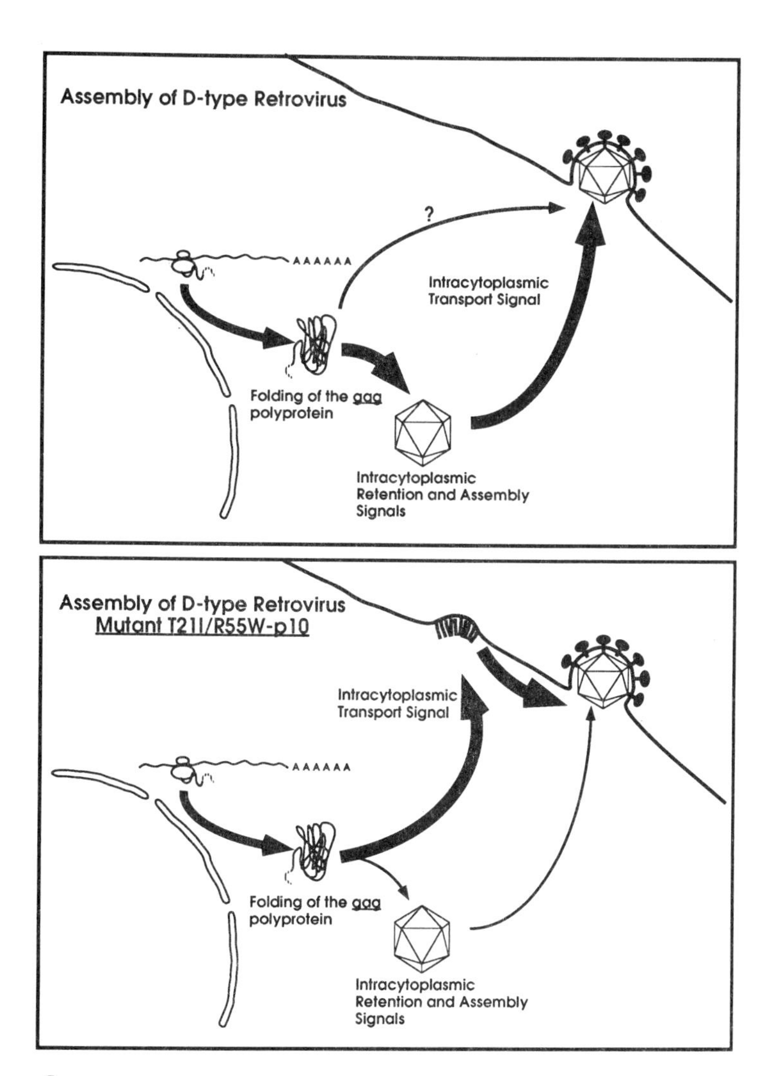

Figure 3.
Proposed model for assembly of intracytoplasmic capsid and transport to the membrane in D-type retrovirus replication. The interpretation of this figure is presented in the text.

A native MA protein structure appears to be essential for *gag* precursor protein stability since a variety of deletion and point mutations, all of which result in the loss of proline residues, destabilize the *gag*-gene product and block capsid assembly. Obviously since p10 is part of a larger precursor, it cannot be ruled out that these mutations have more global effects on precursor structure but equivalent mutations in other *gag* components do not result in reduced polyprotein stability.

Other mutations within the MA coding region affect the rate of capsid assembly and the rate of intracytoplasmic capsid transport, confirming a multi-functional role for the matrix protein in the multi-step process of virus replication - although it should be noted that such functional domains might in fact reside in other regions of the *gag* precursor. Perhaps the most surprising result of these studies is that modification of two amino acid residues within p10 can alter the site of capsid assembly from the cytoplasm to the plasma membrane - effectively converting a D-type virus to a C-type virus. This result suggests that the major difference between the *gag* precursor molecules of these two virus groups is in the dominant transport signals that operate after precursor synthesis. We postulate that in D-type (and presumably B-type) *gag* gene products there exist (at least) two transport signals; a dominant signal that directs the precursor to a cytoplasmic location and specifies retention at that site, and a second signal, functional only after capsid assembly has occurred, that directs them to the plasma membrane (Figure 3). In the *gag*-gene products of C-type viruses and the mutant T21I/R55W-p10, we would postulate that the dominant retention signal is missing or non-functional, and that the membrane-directing signal now dominates. The transport of these individual molecules to specific sites (myristic acid receptors ?) on the plasma membrane would allow self-assembly, virion budding and precursor processing to occur. While such a model is clearly an over simplification of a complex series of molecular interactions and does not take into account necessary virion RNA/protein interactions, it sets the stage for additional mutational and chimeric protein studies that should allow dissection of this critical stage in retrovirus replication.

REFERENCES

Barker, C. S., Pickel, J., Tainsky, M. and Hunter, E. *Virology*, **153**, 201-214, (1986).

Barker, C. S., Wills, J. W., Bradac, J. A. and Hunter, E. *Virology*, **142**, 223-240,(1985).

Bradac, J. and Hunter, E. *Virology*, **138**, 260-275, (1984).

Bradac, J. and Hunter, E. *Virology*, **150**, 503-508, (1986).

Cardiff, R. D., Puentes, M. J., Young, L. J. T., Smith, G. H., Teramoto, Y. A., Altrock,B. W. and Pratt, T. S. *Virology*, **85**, 157-167, (1978).

Gebhardt, A. J., Bosch, V., Ziemiecki, A. and Friis, R. R. *J. Mol. Biol.*, **174**, 297-317, (1984).

Gelderblom, H. R., Hausmann, E. H. S., Ozel, M., Pauli, G. and Koch, M. A. *Virology*, **156**, 171-176, (1987).

Henderson, L. E., Sowder, R., Smythers, G., Benveniste, R. E. and Oroszlan, S. *J. Virol.*, **55**, 778-787, (1985).

Hogle, J. M., Chow, M. and Filman, D. J. *Science*, **229**, 1358-1363, (1985).
Hunter, E. In 'Membrane insertion and transport of viral glycoproteins: a mutational analysis. Protein transfer and organelle biogenesis', (Das, R. and Robbins, P. ed.) pp. 109-158, Academic Press, (1988).
Kawai, S. and Hanafusa, H. *Proc. Natl. Acad. Sci. USA.*, **70**, 3493-3497, (1973).
Linial, M., Fenno, J., Burnette, W. N. and Rohrschneider, L. *J. Virol.*, **36**, 280-290,(1980).
Marcus, S. L., W., S., Racevskis, J. and Sarkar, N. H. *Virology*, **86**, 398-412,(1978).
Marx, P. A., Munn, R. J. and Joy, K. I. *Lab Invest.*, **58**, 112-118, (1988).
Montelaro, R. C., Sullivan, S. J. and Bolognesi, D. P. *Virology*, **84**, 19-31, (1978).
Nermut, M. V., Frank, HP and Schafer, W. *Virology*, **49**, 345-358, (1972).
Palade, G. *Science*, **189**, 347-358, (1975).
Paul, A. V., Schultz, A., Pincus, S. E., Oroszlan, S. and Wimmer, E. *Proc. Natl.Acad. Sci. USA.*, **84**, 7827-7831, (1987).
Pepinsky, R. B. and Vogt, V. M. *J. Mol. Biol.*, **131**, 819-837, (1979).
Rhee, S. S. and Hunter, E. *J. Virol.*, **61**, 1045-1053, (1987).
Rossman, M. G., Arnold, E., Erickson, J. W., Frankenberger, E. A., Griffith, J. P.,Hecht, H. J., Johnson, J. E., Kamer, G., Luo, M., Mosser, A. G., Rueckert, R. R.,Sherry, B. and Vriend, G. *Nature*, **317**, 145-153, (1985).
Schmidt, M., Muller, H., Schmidt, M. F. G. and Rott, R. *J. Virol.*, **63**, 429-431,(1989).
Schultz, A. M. and Oroszlan, S. *J. Virol.*, **46**, 355-361, (1983).
Sonigo, P., Barker, C., Hunter, E. and Wain-Hobson, S. *Cell*, **45**, 375-385, (1986).
Voynow, S. L. and Coffin, J. M. *J. Virol.*, **55**, 79-85, (1985).

13
The Action of Retroviral Protease in Various Phases of Virus Replication

M. M. Roberts and S. Oroszlan

The retroviral genome consists of three genes that are essential for replication. The *gag* gene encodes the internal structural proteins: the nucleocapsid (NC) protein which packages the viral RNA, the capsid (CA) protein which encases the nucleoprotein complex and the matrix (MA) protein which forms an inner leaflet structure between the viral capsid and lipid bilayer. The *pol* gene encodes enzymes essential for replication. These are the protease (PR), reverse transcriptase (RT) and integrase (IN). The *env* gene encodes the surface glycoprotein (SU) and transmembrane protein (TM) associated with the lipid bilayer. (The two letter protein nomenclature is that of Leis *et al.*, 1988). Viral assembly starts with the association of the genomic RNA with the *gag*- and the *gag-pol*-polyprotein, which are the primary translational products of the genome size mRNA. These complexes associate with the TM protein at the cell membrane initiating the 'budding' process resulting in the formation of extracellular immature viral particles.

At this stage, the function of the PR is to cleave the polyproteins into the functional protein components. This cleavage is essential for the production of infectious progeny virus (Katoh *et al.*, 1985). During the maturation process, the core of the virus is collapsed into a capsid that is either icosahedral or cone-shaped as in the case of the lentiviruses (Gonda *et al.*, 1978).

The initiation of PR cleavage in the extracellular immature virion is believed to be an autocatalytic process. For the protease to be active it is required to be in the dimeric form. This has been initially suggested by modelling to the structure of cellular aspartic proteases (Pearl and Taylor, 1987). The dimeric form has been confirmed by X-ray crystallography (Miller *et al.*, 1989; Navia *et al.*, 1989; Wlodawer *et al.*, 1989; Lapatto *et al.*, 1989) and by gel filtration of the active enzyme (Meek *et al.*, 1989). The first cleavage would be generated by the dimeric polyprotein precursor. The PR would be released by cleavage in *cis* (intramolecularly), or *trans* (intermolecular cleavage of neighboring precursor). Much remains to be learned of the precise mechanism(s) of PR activation, the control of proteolysis and the order of fragmentation during virus maturation. Extensive protein sequencing carried out primarily in our laboratory has identified maturation cleavage sites for a large number of retroviruses (for review see Oroszlan and Luftig 1990). The locations and

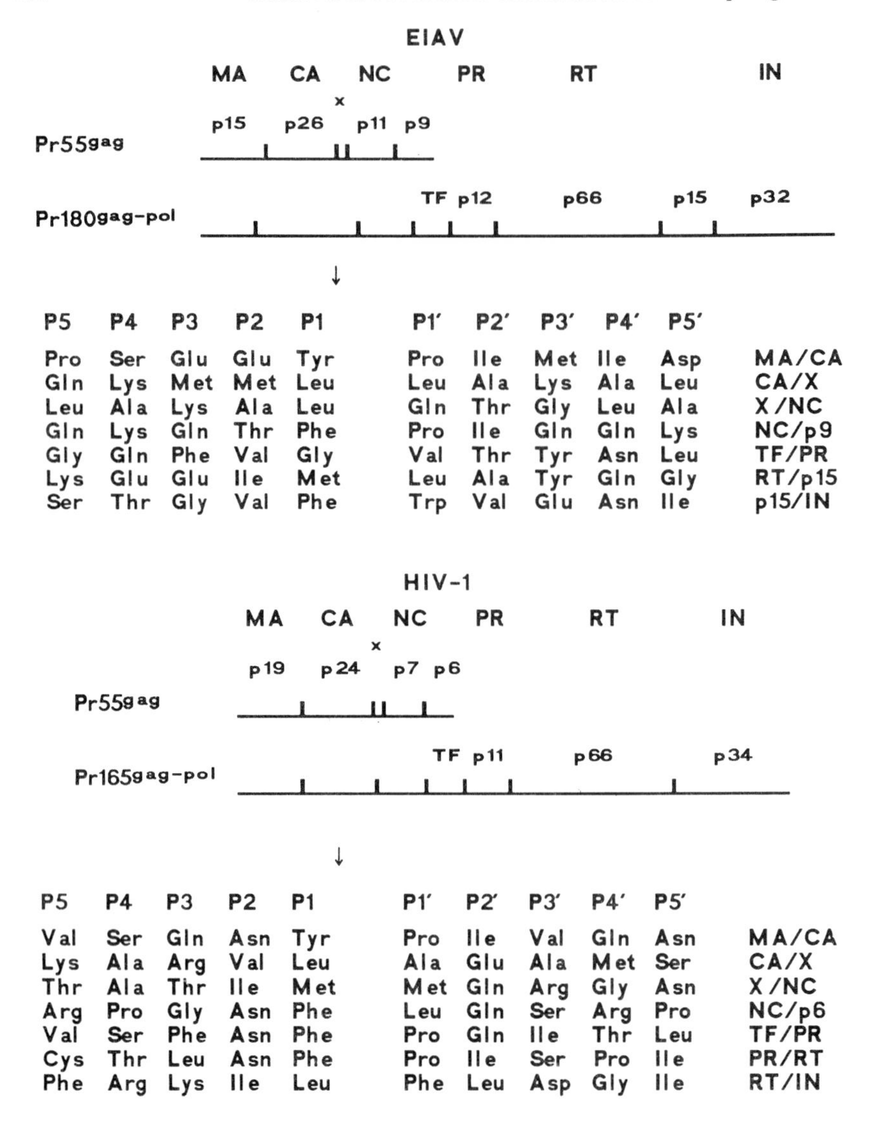

EIAV

P5	P4	P3	P2	P1	P1′	P2′	P3′	P4′	P5′	
Pro	Ser	Glu	Glu	Tyr	Pro	Ile	Met	Ile	Asp	MA/CA
Gln	Lys	Met	Met	Leu	Leu	Ala	Lys	Ala	Leu	CA/X
Leu	Ala	Lys	Ala	Leu	Gln	Thr	Gly	Leu	Ala	X/NC
Gln	Lys	Gln	Thr	Phe	Pro	Ile	Gln	Gln	Lys	NC/p9
Gly	Gln	Phe	Val	Gly	Val	Thr	Tyr	Asn	Leu	TF/PR
Lys	Glu	Glu	Ile	Met	Leu	Ala	Tyr	Gln	Gly	RT/p15
Ser	Thr	Gly	Val	Phe	Trp	Val	Glu	Asn	Ile	p15/IN

HIV-1

P5	P4	P3	P2	P1	P1′	P2′	P3′	P4′	P5′	
Val	Ser	Gln	Asn	Tyr	Pro	Ile	Val	Gln	Asn	MA/CA
Lys	Ala	Arg	Val	Leu	Ala	Glu	Ala	Met	Ser	CA/X
Thr	Ala	Thr	Ile	Met	Met	Gln	Arg	Gly	Asn	X/NC
Arg	Pro	Gly	Asn	Phe	Leu	Gln	Ser	Arg	Pro	NC/p6
Val	Ser	Phe	Asn	Phe	Pro	Gln	Ile	Thr	Leu	TF/PR
Cys	Thr	Leu	Asn	Phe	Pro	Ile	Ser	Pro	Ile	PR/RT
Phe	Arg	Lys	Ile	Leu	Phe	Leu	Asp	Gly	Ile	RT/IN

Figure 1.
PR cleavage sites in the *gag* and *gag-pol* polyproteins of EIAV and HIV-1. The cleavage sites in the *gag* polyprotein of EIAV have been previously published and are taken from Henderson *et al.*, 1987 and Stephens *et al.*, 1986. The cleavage sites in the *pol* domain were determined by sequence analysis of PR, p15 and p32 purified from capsids (unpublished results). For HIV-1 the cleavage site sequences shown are taken from Henderson *et al.*, (1988b) and Ratner *et al.*, 1985. TF, transframe protein. X, excised small peptides.

the nature of the scissile bonds, and the surrounding amino acid sequences for equine infectious anemia virus (EIAV) and human immunodeficiency virus (HIV) are given in Figure 1. It can be seen that there is a limited conservation and a substantial variation in amino acid residues at the scissile peptide bond and in the neighboring sequence. Indeed it is difficult to define a consensus sequence for cleavage.

In addition to the known maturation cleavages, our recent biochemical studies with intact capsids purified from whole EIAV have revealed newly-identified post maturation cleavage sites in the NC protein that are believed to be important for viral replication during the early stages of infection. EIAV and HIV are both lentiviruses, with a high degree of sequence homology (Stephens *et al.*, 1986) and similarity in virus morphology (Gonda *et al.*, 1978). This makes EIAV a good model for HIV, the causative agent of aquired immunodeficiency syndrome (AIDS).

Capsids of EIAV were prepared as described from detergent treatment of whole virus followed by two cycles of rate-zonal centrifugation through a ficoll density gradient (Roberts and Oroszlan, 1989). These cone-shaped particles are 60 nm at the wide end, 25 nm at the narrow end, and 120 nm long, as measured from electron micrographs. Microinjection of these capsids into uninfected cells has shown that they contain all the components required to initiate an infection (Oroszlan *et al.*, 1990). All the *pol*-encoded proteins are present in the capsid together with the viral RNA. Of the *gag*-encoded structural proteins, only the NC (p11) and CA (p26) are present in a 1:1 molar ratio. During incubation of capsids at room temperature or 37°C at pH 7.6 in 10 mM Tris-HCl 1 mM EDTA (TE) buffer, cleavage of the NC protein into smaller fragments takes place. The peptide fragments from the NC cleavage have been separated by RP-HPLC and the chemical composition of each has been determined by sequencing and amino acid analysis (Roberts *et al.*, 1990).

```
1                 10                20
Q T G L A G P F K G G A L K G G P L K A

21    ↓           30
A Q T C Y N C G K P G H L S S Q C R A

40                50                60
P K V C F K C K Q P G H F S K Q C R S V P
        ↑
61                70          76
K N G K Q G A Q G R P Q K Q T F
```

Figure 2.

Locations of the cleavage sites in the EIAV NC proteins (p11) sequence shown in the single letter code. Identities in the internally duplicated regions are shown in bold. Cleavage occurs after the first Cys residue in each conserved domain.

The cleavage sites are located after the first Cys residue within both conserved regions of the sequence (Figure 2). In the absence of reducing agent, cleavage occurs primarily at the Cys(24)-Tyr(25) bond with only limited cleavage at the Cys(43)-Phe(44) bond. Under these conditions, the larger cleavage fragment, a 52 residue peptide, was visible as a p6 band after SDS-PAGE (Roberts and Oroszlan 1989). In the presence of 10 mM DTT, the Cys(43)-Phe(44) cleavage is complete and proceeds after the Cys(24)-Tyr(25) cleavage. On SDS-PAGE this has been seen as the disappearance of the p11 band, through p6 into a diffuse band which contained three fragments from the NC cleavage. The viral aspartic protease has been identified as the enzyme responsible for these cleavages. Pepstatin A, a general aspartic protease inhibitor and specific inhibitors of retroviral PR have been shown to inhibit the processing of the NC protein (Roberts and Oroszlan 1989, Roberts *et al.*, 1990). The endogenous protease cleavages have been found to occur in the viral capsid *in vitro* only when EDTA was present in the buffer used for preparation from whole virus and during incubation. When 10 mM Tris HCl at pH 7.6 is used without EDTA, no cleavage was detected.

	P5	P4	P3	P2	P1	P1′	P2′	P3′	P4′	P5′
EIAV	Ala	Ala	Gln	Thr	Cys	Tyr	Asn	Cys	Gly	Lys
	Ala	Pro	Lys	Val	Cys	Phe	Lys	Cys	Lys	Gln
HIV	Lys	Met	Val	Lys	Cys	Phe	Asn	Cys	Gly	Lys
	Arg	Lys	Lys	Gly	Cys	Trp	Lys	Cys	Gly	Lys
SIV	Lys	Pro	Ile	Lys	Cys	Trp	Asn	Cys	Gly	Lys
	Arg	Arg	Gln	Gly	Cys	Trp	Lys	Cys	Gly	Lys
BLV	Pro	Pro	Gly	Pro	Cys	Tyr	Arg	Cys	Leu	Lys
	Pro	Pro	Gly	Pro	Cys	Pro	Ile	Cys	Lys	Asp
HTLV-1	Pro	Asn	Gln	Pro	Cys	Phe	Arg	Cys	Gly	Lys
	Pro	Pro	Gly	Pro	Cys	Pro	Leu	Cys	Gln	Asp
RSV	Ser	Arg	Gly	Leu	Cys	Tyr	Thr	Cys	Gly	Ser
	Ser	Arg	Glu	Arg	Cys	Gln	Leu	Cys	Asn	Gly
MuLV	Asp	Arg	Asp	Gln	Cys	Ala	Tyr	Cys	Lys	Glu
MMTV	Glu	Gly	Pro	Val	Cys	Phe	Ser	Cys	Gly	Lys
	Pro	Pro	Gly	Leu	Cys	Pro	Arg	Cys	Lys	Lys

Figure 3.

Sequence alignment of the determined and potential cleavage sites within the conserved domains of the NC proteins from a selection of retroviruses. EIAV – (Stephens *et al.*, 1986; Henderson *et al.*, 1987); HIV – (Ratner *et al.*, 1985; Henderson *et al.*, 1988b); SIV – (Henderson *et al.*, 1988a); BLV – (Copeland *et al.*, 1983a); HTLV-1 – (Copeland *et al.*, 1983b); RSV and MuLV – (Henderson *et al.*, 1981); MMTV – (Hizi *et al.*, 1987).

An examination of the EIAV NC protein (p11) sequence (Figure 2) shows that the peptide bonds are cleaved at Cys residues that are followed by large hydrophobic residues (Tyr and Phe) rather than small, or hydrophilic residues (Gly, Arg, Lys). Similar sites can be located in the conserved regions of the NC proteins of other retroviruses as shown in Figure 3. As for EIAV in most cases there is a large hydrophobic residue following the first Cys residue of each conserved ($CX_2CX_4HX_4C$) region, where the cleavages are predicted to occur. In fact, virus suspensions of HIV, simian immunodeficiency virus (SIV) and murine leukemia virus (MuLV) when treated with the preparation protocol for EIAV capsids, did show cleavage of the NC protein after incubation (unpublished data). The observed cleavages reveal a new class of cleavage sites by the viral PR in terms of their amino acid sequence that are highly conserved but atypical of any of the sites maturation cleavage in that Cys occurs in the P1 position.

As for maturation cleavages, there could be a conformational requirement for cleavage of the NC protein. Recently, a high resolution NMR structure was determined for an 18-residue peptide containing the sequence of the first conserved region ($CX_2CX_4HX_4C$) of the HIV NC protein bound to zinc (Summers *et al.*, 1990). The Zn atom is tetrahedrally coordinated to three Cys and one His residue. The first and second Cys residues are part of a tight four residue turn that is believed to be important in viral RNA recognition and binding. It is this exposed site which is cleaved by the viral PR. However, in order for this site to become the substrate, it would have to assume the extended conformation required for cleavage, and bound zinc, if present, would have to be displaced. The requirement of EDTA for the *in vitro* cleavage would support this hypothesis. We propose here that this cleavage also occurs *in vivo* upon the release of the capsid into the cytoplasm following membrane fusion during entry of virus into the uninfected cell. With the NC protein cleaved, the RNA would be uncoated and free to proceed to the next essential steps towards infection (reverse transcription/integration, see Figure 4).

To explore the effect of the NC cleavage on the endogenous reverse transcription of the viral RNA, the capsids were preincubated to allow NC cleavage to occur before the RT reaction was started. Reactions were performed in the presence and absence of protease inhibitor. In the absence of pepstatin A, slightly higher TCA-precipitable ^3H-DNA counts were obtained (Table 1), and particles of 80 nm in size were produced as observed by EM (Figure 5). In the presence of pepstatin A, or in RT reactions without preincubation particles of this size range with uniform shape were not observed; only smaller (20–40 nm) particles were seen. The morphology of the 80 nm particles is not certain, though they are believed to be icosahedral.

Particles from the RT reaction mixture cosedimented with ^3H-labelled DNA and banded at an equilibrium density of 1.34 g cm^{-3} in CsCl. All the counts were recoverable by TCA precipitation. Analysis by SDS-PAGE showed that the banded product contains the CA protein as the major component and the IN protein as a minor component.

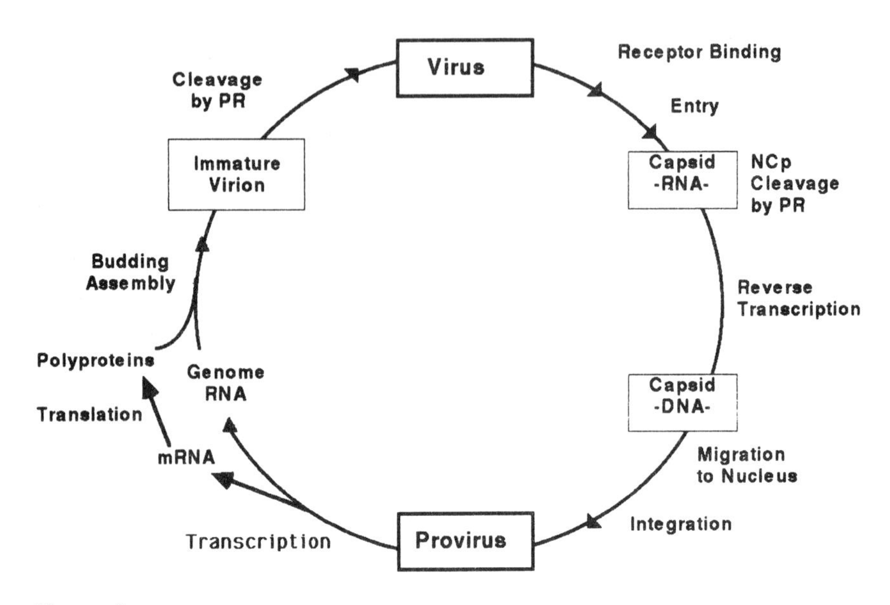

Figure 4.
An outline of the retroviral life-cycle adapted from the sketch by Varmus and Brown (1989) to illustrate the involvement of PR in both the early and late stages of virus replication.

Table 1.
RT Activity of RNA Capsids [a]

	Samples Preincubated 4h, 37xC in pH 7.6 TE	
Incubation Time	PR Inhibitor —	PR Inhibitor +
0 h	1.5 x 10	1.6 x 10
	1.5×10^4	1.8×10^4
5 h	8.6×10^5 [b]	7.6×10^5
	8.6×10^5	7.3×10^5

[a] Measured as TCA-precipitable cpm in RT cocktail. ^{3}H-TTP 0.08 μCi/μl (60 Ci/mmol); d(NTP) (N = A,G,C), 238 μM; Tris HCl, 80 mM; NaCl, 36 mM; Triton-X100, 0.036%; $MgCl_2$, 6 mM; DTT, 8 mM at pH 7.4 in pH 7.6 TE.

[b] Icosahedral particles (φ = 80 nm, ρ = 1.34 g cm^{-3}) produced under these conditions.

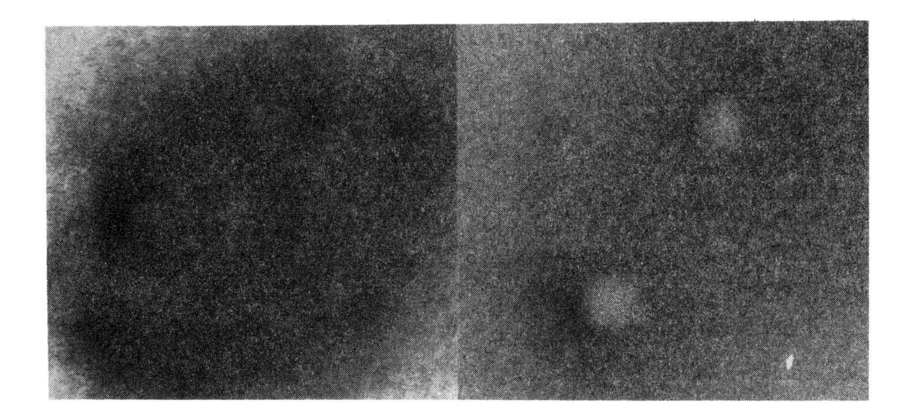

Figure 5.
Negative-stained electron micrographs (x90,000). Purified EIAV capsids which contain RNA (left frame). DNA containing capsids (right frame) obtained after the cone shaped RNA capsids were first incubated to allow PR cleavage and then incubated for RT reaction as in Table 1 in the absence of PR inhibitor.

Recently 160S particles containing DNA, p30 (CA) and IN have been isolated from the cytoplasm of cells freshly infected with MuLV (Varmus and Brown, 1989). The purified cytoplasmic particles have been shown to be integration competent *in vitro*. It is suggested that the particles seen by EM (Figure 5 right frame) represent nucleoprotein complexes similar to those occurring *in vivo*. If this is so, then the NC cleavage is required not only to release the RNA for reverse transcription, but also to allow particle 'metamorphosis' to occur, during which full length linear double stranded DNA can be correctly packaged for integration.

The chemistry of the NC protein cleavage process is now well defined (Roberts *et al.*, 1990). The observed biochemical processes that occur during the *in vitro* reverse transcription reactions with capsids need to be thoroughly pursued with further experiments including extension to *in vivo* studies in order to define a role, if any, for the protease during the early stages of the retroviral life-cycle. It is conceivable that protease inhibitors could block the biochemical events required for integration of viral DNA (see Fig. 4). Highly specific and potent inhibitors of the viral protease would prevent infection at this early stage of the virus life cycle before the provirus is formed and could be used as prophylactic agents against HIV and AIDS.

ACKNOWLEDGEMENTS.

We thank S. Lockhart, S. Bricker, and M.B. Cassell for excellent technical assistance. We are also grateful to C. Hixson and M. Bowers for amino acid analysis, Y.D. Kim and T.D. Copeland for protein sequencing, R. Sowder for his advice on the RP-HPLC separation, K. Nagashima for the electron micrographs, and C. Rhoderick for typing the manuscript.

Research sponsored by the National Cancer Institute, DHHS under contract No. NO1-CO-74101 with BRI. The contents of this publication do not necessarily reflect the views or policies of the Department of Health and Human Services, nor does mention of trade names, commercial products, or organizations imply endorsement by the U.S. Government.

REFERENCES

Copeland,T.D., Morgan,M.A. and Oroszlan,S. *FEBS Lett.*, **156**, 37-40, (1983).

Copeland,T.D., Oroszlan,S., Kalyanaraman,V.S., Sarngadharan,M.G. and Gallo,R.C. *FEBS Lett.*, **162**, 390-395, (1983).

Gonda,M., Charman,H.P., Walker,J.L., and Coggins, L. *Am. J. Vet. Res.*, **39**, 731-740, (1978).

Henderson,L.E., Benveniste,R., Sowder,R. Copeland,T.D. Schultz,A.M and Oroszlan,S. *J. Virol.*, **62**, 2587-2595, (1988).

Henderson,L.E., Copeland,T.D., Sowder,R.C., Schultz,A.M., and Oroszlan,S. in 'Human Retroviruses, Cancer and AIDS: Approaches to Prevention and Therapy', (Bolognesi, D. ed), pp.135-147, Alan R. Liss, Inc., (1988).

Henderson,L.E., Copeland,T.D., Sowder,R.C., Smythers,G.W. and Oroszlan,S. *J. Biol. Chem.*, **256**, 8400-8406, (1981).

Henderson,L.E., Sowder,R.C., Smythers,G.W. and Oroszlan, S. *J. Virol.*, **61**, 1116-1124, (1987).

Hizi,A., Henderson,L.E., Copeland,T.D., Sowder,R.C., Hixson,C.V. and Oroszlan,S. *Proc. Natl. Acad. Sci. USA*, **84**, 7041-7045, (1987).

Katoh,I., Yoshinaka,Y., Rein,A., Shibuya,M., Odaka,T. and Oroszlan,S. *Virology*, **145**, 280-292, (1985).

Lapatto,R., Blundell,T., Hemmings,A., Overington,J., Wilderspin,A. Wood,S. Merson,J.R., Whittle,P.J., Danley,D.E., Geoghegan,K.F., Hawrylik,S.J., Lee,S.E., Scheld,K.G. and Hobart,P.M. *Nature*, **342**, 299-302, (1989).

Leis,J., Baltimore,D., Bishop,J.M.., Coffin,J., Fleisner,E., Goff,S.P., Oroszlan,S., Robinson,H., Skalka,A.M., Temin,H.M. and Vogt, V. *J. Virol.*, **62**, 1804-1809, (1988).

Meek,T.D., Dayton,B.D., Metcalf,B.W., Dreyer,G.B., Strickler,J.E., Gorniak,J.G. Rosenberg,M., Moore,M.L. Magaard,V.W. and Debouck,C. *Proc. Natl. Acad. Sci. USA*, **86**, 1841-1845, (1989).

Miller,M., Jaskolski,M., Rao,J.K.M., Leis,J. and Wlodawer,A. *Nature*, **337**, 576-579, (1989).

Navia,M.A., Fitzgerald,P.M.D., McKeever,B.M. Leu,C.-T., Heimbach,J.C, Herber,W.K., Sigal,I.S. Darke,P.L. and Springer,J.P. *Nature*, **337**, 615-620, (1989).

Oroszlan,S., Boyd,A., Cassell,M.B. and Roberts,M.M. in preparation, (1990).

Oroszlan,S. and Luftig,R.B. in 'Current Topics in Microbiology and Immunology', (Swanstrom,R. and Vogt,P.K. eds), Springer-Verlag, (1990), in press.

Pearl,L.H. and Taylor,W.R. *Nature*, **329**, 351–354, (1987).

Ratner,L., Haseltine,W., Patarca,R., Livak,J., Starcich,B., Josephs,S.F., Doran,E.R. and Rafalski,J.A. *Nature*, **313**, 277–281, (1985).

Roberts,M.M., Copeland,T.D., and Oroszlan,S. submitted (1990).

Roberts,M.M. and Oroszlan,S. *Biochem. Biophys. Res. Commun.*, **160**,

Stephens,R.M., Casey,J.W. and Rice,N.R. *Science*, **231**, 589–594, (1986).

Summers,M.F., South,T.L., Kim,B. and Hare,D.R. *Biochemistry*, **29**, 329–340, (1990).

Varmus,H. and Brown,P. in 'Mobile DNA' (Berg, D.E. and Howe, M.M., eds), pp. 53–108, American Society for Microbiology, (1989).

Wlodawer,A., Miller,M., Jaskolski,M., Sathyanarayana,B.K., Baldwin,E., Weber,I.T., Selk,L.M., Clawson,L., Schneider,J. and Kent,S.B.H. *Science*, **245**, 616–621, (1989).

14
Terminal Stages of Retrovirus Morphogenesis

Ronald Luftig, Kazuyoshi Ikuta, Ming Bu and Peter Calkins

Our current model for describing the terminal stages of murine leukemia virus (MLV) morphogenesis is presented in Figure 1. It is derived from biochemical and ultrastructural results of ours and other laboratories, over the past 15–20 years (Luftig and Kilham, 1971; Bolognesi, Luftig and Shaper, 1973; Yoshinaka and Luftig, 1977a,b; Jamjoom, Naso and Arlinghaus, 1976; Yoshinaka *et al.*, 1985; Katoh *et al.*, 1985; Traktman and Baltimore, 1982; Crawford and Goff, 1985; Ikuta and Luftig, 1988). A critical role is noted for polyproteins $Pr65^{gag}$ and $Pr180^{gag\text{-}pol}$, as well as the viral encoded protease (PR) in the final stages of assembly.

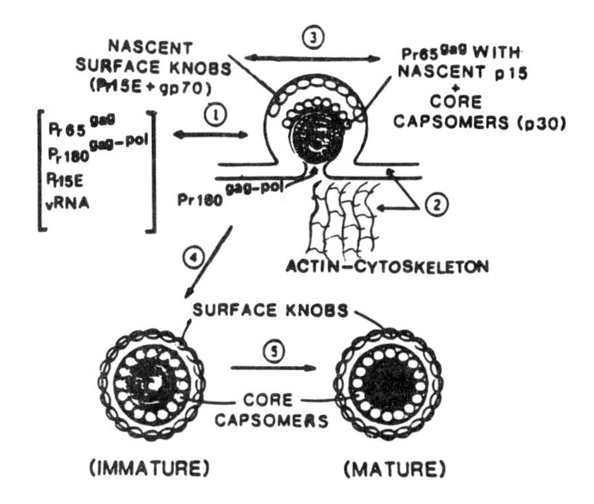

Figure 1.
Model of MLV morphogenesis is divided into five steps. *Step 1*: Formation of the budding particle from uncleaved polyproteins and genomic RNA, *Step 2*: Association of the viral bud with cellular cytoskeleton or membrane elements, *Step 3*: Association of *gag* and @env@ encoded determinants at the cell membrane, *Step 4*: Activation of PR cleavage during 'bud' release and *Step 5*: Conversion of 'immature' to 'mature' infectious particles.

RESULTS AND DISCUSSION

There are several interesting features to emphasize in the model of Figure 1. *First*, we estimate that ≈1500 precursor polyprotein molecules are stably associated with two 35s RNA genome equivalents in an 'immature' particle. This is based on utilizing a model for alphaviruses to estimate the number of MLV p30 globular molecules on the surface of a spherical mature core (see Appendix, Fig. 6). *Second*, since $Pr180^{gag-pol}$ is only present at about 5% the level of $Pr65^{gag}$ in MLV-infected cells and it is the precursor to RT of which there are ≈60 molecules/virions (Panet, Baltimore and Hanafusa, 1975), it is structurally consistent to locate 5 molecules of $Pr180^{gag-pol}$ at each of the icosahedral vertices.

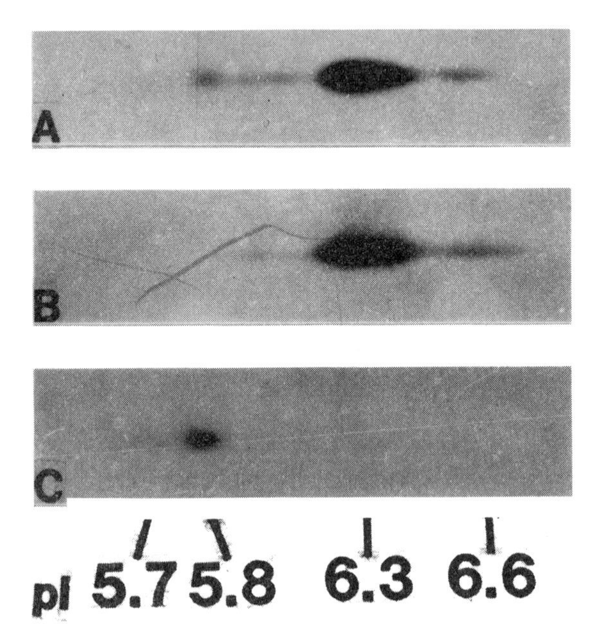

Figure 2.
2D SDS-PAGE analysis of A) a mixed virion sample, one part which was labeled *in vitro* with γ-^{32}P ATP and the other *in vivo* with ^{35}S-methionine prior to immunoprecipitation with anti-p30 sera. B) the *in vivo* ^{35}S-methionine labeled clone 2 virions, immunoprecipitated with anti-p30 sera, and C) *in vitro* γ-^{32}P ATP labeled clone 2 virions. Clone 2 cells, chronically infected with MLV had been grown in roller bottles and virion fractions were purified from the culture fluid. *In vitro* phosphorylation occurred in a standard mixture containing 1% NP-40, 20mM Tris-HCl, pH 8.0, 5mM $MgCl_2$, 10mM DTT and 10μm ATP incubated at 37°C for 20 min. We note in C) that phosphorylation occurs *in vitro* for the same isoform as was noted when ^{32}PPi labeling was done *in vivo*, prior to isolation of virions (Ikuta and Luftig, 1988). As noted in the mixing experiment of panel A) this spot does not overlap any of the major ^{35}S-methionine p30 isoforms (pI: 6.0, 6.1, 6.3, 6.6).

Third, PR activation is presumed to occur at only one vertex, the closure site of a budding 'immature' particle. Based on our experimental finding that <1% of p30 molecules are phosphorylated (Ikuta and Luftig, 1988) and the calculation that 0.33% of all p30 in a virion (derived from $Pr65^{gag}$, as well as $Pr180^{gag-pol}$ would be at a single vertex of an icosahedron with triangulation number $T=25$ (see Appendix, Figure 6), I suggest that p30 phosphorylation of $Pr180^{gag-pol}$ at the bud closure site is sufficient to trigger a change in charge and/or pH, leading to PR activation by autocatalysis of two $Pr180^{gag-pol}$ molecules. We have identified the phosphorylated p30 tryptic peptide from *in vivo* ^{32}PPi labeling experiments and believe that one or more residues of a Ser-Ser-Ser triplet near the NH_2 terminus of MLV p30 are phosphorylated. It can be seen in Figure 2C that only phosphorylation of the specific p30 component with pI: 5.8 occurs on 2D gels.

Evidence to support the idea that the bud closure site is unique, comes from the HVEM observation of budding particles. As seen in the insets of Figure 3A, a necklace of protein subunits surrounds the closure site (arrows). Further, for 'mature' particles entering the cell there appears to be an unwinding of similarly sized subunit material at the cell surface, which may play a role in permitting insertion of the 'mature' core into the cell at a single vertex. It is of interest to note that HVEM,

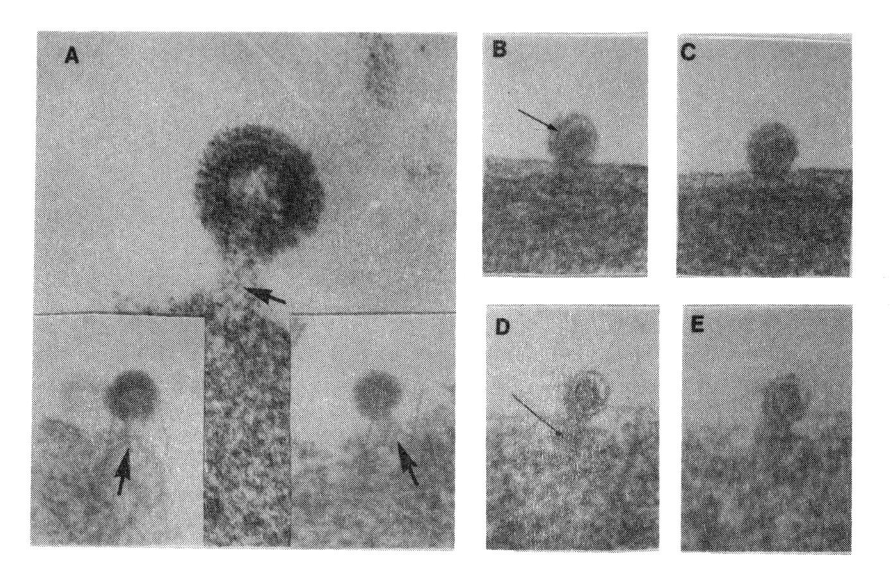

Figure 3.
HVEM of A) 'Immature' budding particles. Arrows on insets point to protein necklace, which can be seen to be made up of subunits on larger photo. B), C) and D),E) are stereo pairs of same 'mature' particle which appears to be entering the cell by a proteinaceous tail, (B,D), which appears external to the hexagonal core (C,E).

which utilizes thick (0.25–0.5 μm) sections and stereo pairs, permits us to see such substructures on 'mature' cores, which in conventional thin section (40–60nm), only appear as hexagonal, darkly-staining structures (Figure 3C, E).

In the final event of maturation, the activated PR is thought to cleave $Pr65^{gag}$, as well as $Pr180^{gag-pol}$ molecules from vertex to vertex in what will become a 'mature' core icosahedron (Yoshinaka and Luftig, 1978). One of the difficulties in studying this process *in vitro* is that only minor amounts of MLV PR can be obtained from virions. MLV PR in contrast to the avian (RSV) *gag* encoded PR is a *gag-pol* encoded, unstable enzyme (Yoshinaka and Luftig, 1980). Over 50mg of MLV is required to obtain small amounts (μg) of MLV PR (Yoshinaka and Luftig, 1977a; Yoshinaka *et al.*, 1985). In order to increase yields, we have cloned the MLV PR gene into a *trp*E vector, and obtained about 1g of PR from 1 liter of cells. This has permitted a partial characterization of some properties, such as pH optimum for the cleavage reaction. As noted in Figure 4, the bacterially produced MLV PR exhibits an optimum at pH 5.5, similar to that of other retroviral proteases. In Figure 4, we note that appearance of both a (*trp*E-PR) 50kD fusion product, and the 14kD PR, occurs only when the MLV PR is active.

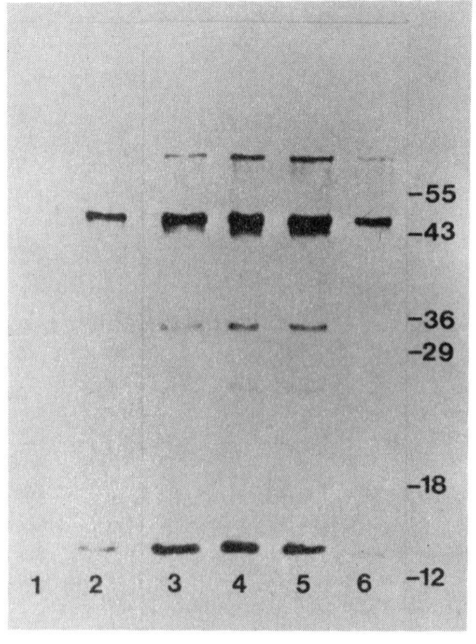

Figure 4.
A partially purified preparation of the MLV PR was obtained from a *trp* E expression system. Lanes 1–6 are blotted with MLV PR antisera, and represent assays performed at pH 4.5, 4.7, 5.2, 5.5, 5.9, 6.4.

In addition to studying the MLV-PR, we have initiated experiments on the HIV-PR, utilizing cleavage of 'immature' MLV $Pr65^{gag}$ containing Gazdar particles as an assay. This particle-based assay allows us to detect potential HIV PR inhibitors in a more native setting. Inhibitors that fail to cleave the $Pr65^{gag}$ polyprotein precursor should fail to convert 'immature' to 'mature' particles *in vivo*, since this is the final stage of retrovir us maturation. We have already shown that the antifungal antibiotic cerulenin is an inhibitor of the MLV PR both *in vivo* and *in vitro* (Katoh, Yoshinaka, and Luftig, 1986; Ikuta and Luftig, 1986). Cerulenin also effectively blocks the HIV PR at about 0.2mM concentration in our *in vitro* as say (Bu, Oroszlan and Luftig, 1989) and Pr^{gag} cleavage *in vivo* with HIV-infected H-9 cells (Pal, Gallo and Sarnagadharan, 1988). Unfortunately, cerulenin itself cannot be used therapeutically, since it is fairly toxic to cells. Thus, together with Drs. Blumenstein, Michejda and Oroszlan at the Frederick Cancer Center, a number of cerulenin analogues have been developed to test in our assay system. One of the most effective ones is (2R-*cis*)-Epoxydodecanoyl-L-proline (#105). In preliminary studies this inhibitor is even more effective when mixed together with pepstatin A (Fig. 5). By mixing the two together at 0.1mM, the inhibition is seen to be greater than either compound alone at 0.1mM. Perhaps, in a synergistic manner the cerulenin analog modifies the HIV PR structure so that for example the flap region is open for greater time intervals, allowing pepstatin-A easier access to the DTG site.

Future experimentation on the role of the MLV PR in assembly will utilize information from the crystal structure map of HIV PR (Navia *et al.*, 1989 ; Wlodawer *et al.*, 1989). Alignment of the MLV and HIV PR primary sequences permits specific regions of the HIV PR structure to be compared with putative regions on the MLV PR (A. Wlodawer and I. Weber - pers. commun.). This has provided us an opportunity to design experiments that test for the functional activity of specific domains on the MLV PR. For example, is there any structural significance to the finding that the monomer of MLV PR is 125aa while that of HIV PR is 99aa, yet both exhibit full PR activity in solution? Deletion mutagenesis of the 18aa COOH tail region of MLV PR should help answer this question. In sumary, we have learned and continue to learn a great deal about the role of the PR in retrovirus morphogenesis from studies of MLV assembly. Now that the HIV PR structure has been crystallized, this should help us design future experiments with the MLV PR.

ACKNOWLEDGEMENT

This research was supported in part by National Cancer Institute Grant 5 RO1-CA-37380 to Dr. Ronald B. Luftig and NIH Training Grant (CA-09482) Fellow ship to Peter Calkins. Appreciation for assistance with the modeling of Figure 6 is given to Dr. Russell Durbin (Rutgers University) and for aid with the HVEM (Figure 3) to Dr. Tom Borg (University of South Carolina). Also, Drs. Steve Oroszlan and Jeff Blumenstein generously donated antisera and the cerulenin analog, respectively.

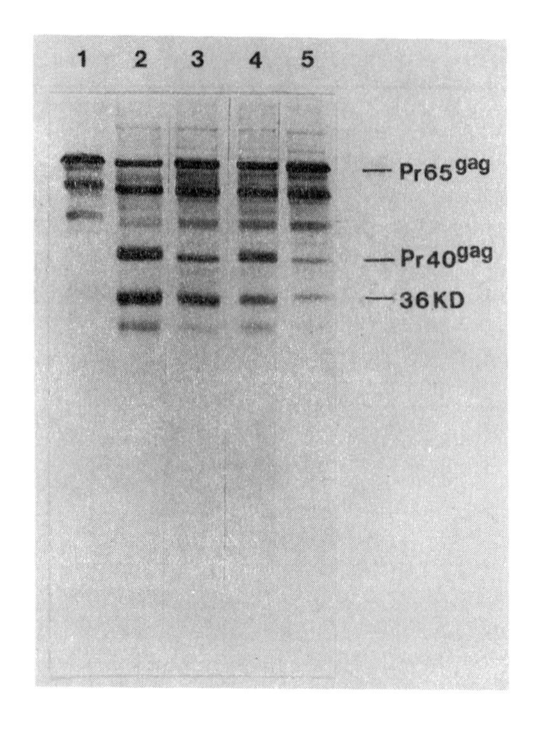

Figure 5.
Enhanced inhibition of MLV $Pr65^{gag}$ cleavage by mixtures of inhibitors. Lane 1 is $Pr65^{gag}$; lane 2 is $Pr65^{gag}$ + HIV PR showing significant cleavage into $Pr40^{gag}$ and a 36k product. Lane 3 is 0.1mM pepstatin A addition to the assay of lane 2; lane 4 is 0.1mM #105 addition; and lane 5 is a mixture of 0.1mM pepstatin A + 0.1mM #105 added to the assay. It can be seen in lane 5, the enhanced inhibition of Pr65 cleavage into $Pr40^{gag}$ and 36K.

REFERENCES

Bolognesi,D.P., Luftig,R.B., and Shaper,J.H. *Virology*, **56**, 549-564, (1973).
Crawford,S. and Goff,S.P. *J. Virol.*, **53**, 899-907, (1985).
Enzmann,P.I. and Weiland,F., *Virology*, **95**, 501-510, (1979).
Ikuta,K. and Luftig,R.B., *J. Virol.*, **62**, 40-46, (1988).
Jamjoom,G.A., Naso,R.B., and Arlinghaus,R.B., *J. Virol.*, **19**, 1054-1072, (1976).
Katoh,I., Yoshinaka,Y., Rein,A., Shibuya,M., Odaka,T., and Oroszlan,S., *Virology*, **145**, 280-292, (1985).
Luftig,R.B., J. *Ultrastruc. Res.*, **20**, 91-102, (1967).
Luftig,R.B. and Kilham,S.S., *Virology*, **46**, 277-297, (1971).
Navia,M.A., Fitzgerald,P.M.D., McKeever,B.M., Leu,C.T., Heimbach,J.C., Herber,W.N., Sigal,I.S., Darke,P.L., and Springer,J.P., *Nature*, **337**, 615-620, (1989).

Nermut,M.V., Frank,H., and Schafer,W., *Virology*, **49**, 345-358, (1972).
Pal,R., Gallo,R.C., Sarnagadharan,M.G., *Proc. Natl. Acad. Sci. USA*, **85**, 9283-9286, (1988).
Panet,A., Baltimore,D., and Hanafusa,H., *J. Virol.*, **16**, 146-152, (1975).
Traktman,R. and Baltimore,D., *J. Virol.*, **44**, 1039-1046, (1982).
Wlodawer,A., Miller,M., Jaskolski,M., Sathyamarayana,B., Baldwin,E., Weber,I.T., Selk,L.M., Clawson,L., Schneider,J., and Kent, S.B.H., *Science*, **245**, 6 16-621, (1989).
Yoshinaka,Y. and Luftig,R.B., *Proc. Natl. Acad. Sci. USA*, **74**, 3467-3470, (1977a).
Yoshinaka,Y. and Luftig,R.B., *Cell*, **12**, 709-719, (1977b).
Yoshinaka,Y. and Luftig,R.B., *J. Gen. Virol.*, **40**, 151-160, (1978).
Yoshinaka,Y. and Luftig,R.B., *Virology*, **111**, 239-250, (1981).
Yoshinaka,Y., Katoh,I., Copeland,T.D., and Oroszlan,S., *Proc. Natl. Acad. Sci. USA*, **82**, 1618-1622, (1985).

APPENDIX

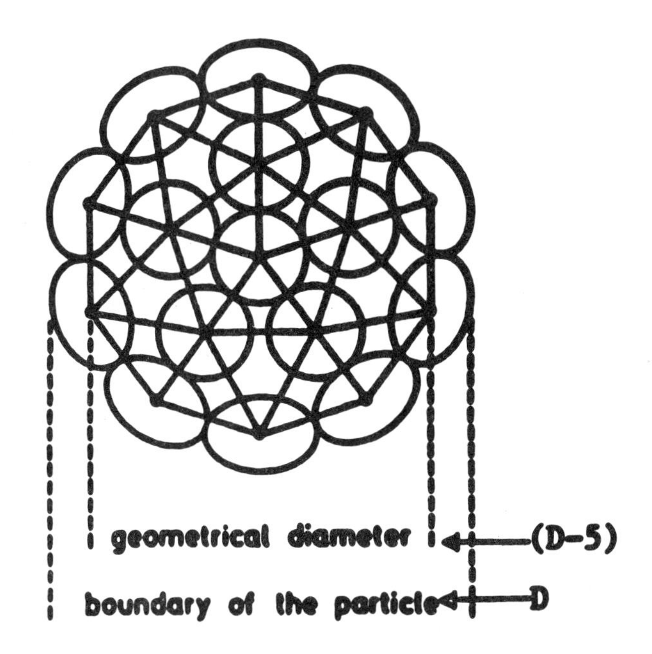

Schematic drawing of a virus as icosahedron.

Figure 6.

As a first approximation to the MLV mature core, we have used this model of an alphavirus adapted from Enzmann and Weiland (1979). In it we substitute for an alphavirus capsomer, an MLV p30 globular protein. The diameter of such a typical protein with Mr of 30kD on the surface is assumed to be 5nm. Then the diameter (D) of the MLV core shell at the point of contact among the shell proteins will be D-5. Further, the area occupied by a single molecule of globuler protein is proportional to r^2 (r = radius of the molecule) and based on physical data from several globular protein, r is proportional to $M^{1/3}$ (M = mass of the molecule). If we now take N to be the number of p30 molecules, and put in proportionality constants, i.e., k, k', k'' , we find that: (i) Nkr^2 = area occupied by all p30 molecules on the shell and (ii) $r = k' M^{1/3}$. Therefore, $Nk(k')^2 M^{2/3}$ = the area of the capsid shell. However, this area can also be considered proportional to $(D-5)^2$ or = $k''(D-5)^2$. Putting all equations together, $Nk(k')^2 M^{2/3} = k''(D-5)^2$ or combining constants, $N = k'''(D-5)^2/M^{2/3}$.

The constant k''' derived from several viruses is 2.7± 0.3. For example, with Adenovirus type 5; N=780 (240 trimers of Mr = 120kd + 60 pentamers of Mr = 70kd); Mr (avg.) of capsid shell protein = 116kd; D=88 nm (based on our EM measurements using the catalase crystal internal marker technique (Luftig, 1967)). Then, $k''' = 780(23.8)/83^2 = 2.7$. Therefore, we can calculate that 'mature' MLV cores with a diameter of 80nm (Nermut, Frank and Schafer, 1972) and containing p30 as the major capsid protein, have $N=2.7(75^2)/(30)^{2/3}$ or 1574 structural units. Then by extrapolation, 'immature' cores which contain the uncleaved precursor polyproteins $Pr65^{gag}$ plus $Pr180^{gag-pol}$, also possess a total of 1574 molecules. As an aside, we note that by utilizing a second estimation technique based on computer graphics of freeze-fractured mature core electron micrographs taken by Nermut *et al.* (1972), we calculate an optimal number of 230 hexamers or 1380 structural units per core. Taken together, both estimates suggest about 1500 structural units, on average cover a mature core. This is most consistent with an icosahedron of $T=25$.

15
Analysis of the Assembly of the HIV Core by Electron Microscope Tomography

Stefan Höglund, Lars-Göran Öfverstedt, Åsa Nilsson, Muhsin Özel, Thorsten Winkel, Ulf Skoglund and Hans Gelderblom

Both the proteolytic processing of the p55 *gag* precursor protein of HIV, myristilation of the p17 (MA) protein, and their morphological rearrangement are necessary to generate structurally mature virus particles (Dahlberg, 1988; Haseltine, 1989). This sequence of events would parallell the observations with e.g. Friend murine oncoviruses, also indicating that a late stage of virus maturation would be accompanied by the production of infectious virus (Katoh *et al.*, 1985).

In HIV the p55 polyprotein precursor is cleaved by the viral proteinase (Orozlan and Luftig, 1989) into the proteins p17 (MA), p24 (CA), p7 (NC) (Leis *et al.*, 1988) and p9/6 *gag* proteins, observed in the cell-released HIV (Henderson *et al.*, 1988; Veronese *et al.*, 1988). The internal proteins of HIV would comprise the structural proteins MA, CA, NC p7 and the enzymes protease p12 (PR), reverse transcriptase p66 (RT), integrase p32 (IN) (Haseltine *et al.*, 1989) in addition to RNA genome (Oroszlan and Luftig, 1989).

Morphogenesis and fine structure of HIV and SIV have been extensively studied by transmission electronmicroscopy (TEM) including the internal protein assembly of the virus (Gelderblom *et al.*, 1987, 1989; Gonda, 1988; Chrystie and Almeida, 1988,1989; Grief *et al.*, 1989). The major virus structural proteins have been localized by immunoelectronmicroscopy (IEM), and the results from both conventional TEM and IEM have depicted a structural model of HIV (Gelderblom *et al.*, 1987, 1989). The viral envelope is studded with SU glycoprotein knobs which are shed spontaneously during virus ageing, and are connected by weak bonds to the gp41 TM protein (Gelderblom *et al.*, 1989). The viral genome is contained within a core consisting of the CA protein, and associated with the NC p7 protein (Veronese *et al.*, 1988). Underneath the lipid bilayer MA is forming a shell, 5 to 7 nm in thickness, probably serving as a morphopoetic factor. Along the core electron-dense material of unknown composition forms lateral bodies (Gelderblom *et al.* 1987). Recently, a regular icosahedral structure of the MA shell has been tested by 3-dimensional computor graphics (Marx *et al.*, 1988).

In this study we extended the dimension of visualization from conventional 2-dimensional TEM projections to 3-dimensional reconstructions of the internal morphology of the virion. By electron microscope tomography (Skoglund and Daneholt, 1986) using a collection of a great number of projections, taken at different tilt angles, a 3-dimensional image of the HIV core would be attainable. The interaction between the structural proteins of the core and the matrix, as deduced from the interior of the sectioned viron, will also be described.

MATERIALS AND METHODS

Cells and virus

HIV-1, strain HTLV-IIIB was propagated in H9 cells and prepared for thin section electron microscopy as detailed earlier (Gelderblom *et al.*, 1987). Freshly, infected cell samples were fixed on day 4 or 5 in 2.5% glutaraldehyde, postfixed in 1% OsO_4 and treated with 1 % tannic acid to improve image contrast in TEM. After agar block enclosure and an additional treatment with 1% uranyl acetate, the cells were embedded in Epon. Routine thin sections, as well as serial thin sections, 40 to 50nm in thickness were cut, mounted on naked 200 mesh copper grids, and poststained with lead citrate (Figure 1). For tomography specimens, between 60 nm and 80 nm in thickness, were prepared to enable the analysis of the entire core. They were mounted on Piloform and carbon-coated, 300 mesh copper grids.

Gold markers

Five nm gold particles were applied to the specimens to produce coordinates for electron microscope tomography (Figure 2). These coordinates are determined to calculate image beam rotation, tilt angle, possible shrinkage and changes of planarity and parallaxes in the whole tilt series.

Electron microscopy

The specimens were analyzed in a ZEISS CEM 902 system, equipped with a goniometer and a liquid nitrogen cooling stage for the specimen. Projections were photographed every 5 (or 3) degrees between -60° and +60°. Controls of the quality of the specimens were recorded at 0° before and after the tilt series.

Electron microscope tomography

About fifteen series of 30 electron micrographs containing the complete set of tilted views, every 5 (or occasionally 3) degrees from +60° to -60°, were digitized in a photoscanner. The proceeding calculations were accomplished using a coherent computor software package. The midpoint of a number of fixpoints, represented by the gold particles previously applied to the specimen, were identified. By this means the parameters of tilt angle, rotation and translation can be estimated through iterative least square method.

a.

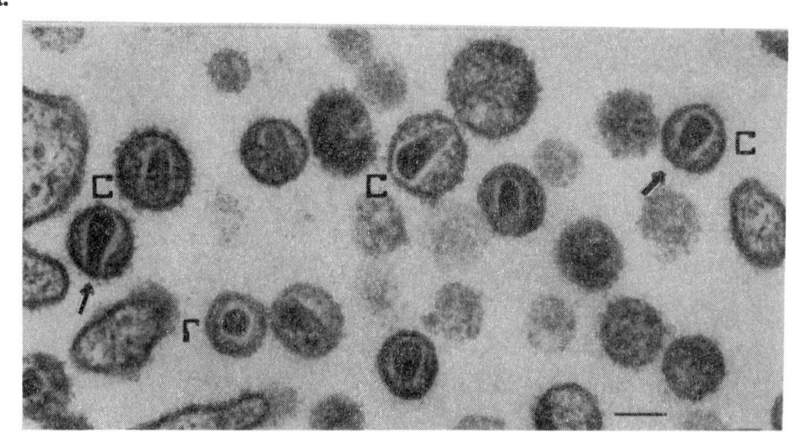

b.

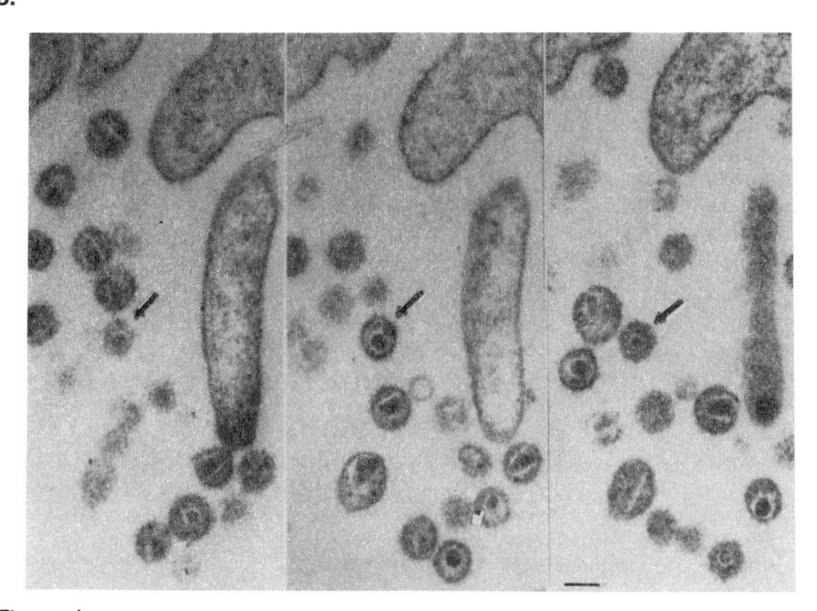

Figure 1.
a) A population of sectioned HIV particles demonstrating various core projections: cone-shaped (**c**) and rounded (**r**). A possible connecting site between core and envelope is indicated by arrow. **b)** TEM of three consecutive serial sections of a virus particle showing a variation in the diameter of the core structure (indicated by arrows). (Bar indicates 100 nm.)

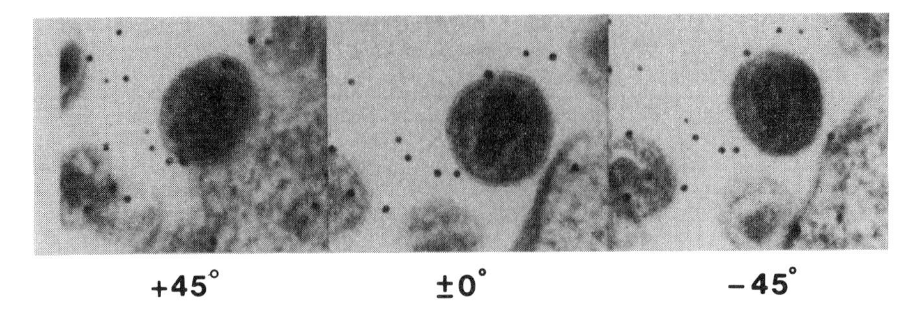

Figure 2.
Gold particles as coordinates for tomography are shown adjacent to an HIV particle at different tilt angles.

In a series of 2-dimensional sections a circular disc was reconstructed by backprojection of the corresponding lines of the micrographs in each single step (Figure 3). The 3-dimensional reconstruction was done by Radius-weighted backprojection after a Fourier transform of the different projections of the particle. Ultimately, several consecutive discs will form a cylinder on which the reconstructed object is contained. The structure model obtained is then being noise filtered to a proper resolution through Fourier filtering: usually a resolution of 3 nm may be obtained on a 35 nm big object.

The completed reconstruction is shown as a number of density maps, corresponding to the reconstructed 2-dimensional sections (Figure 3). The analyzed particle may be presented as a contoured vector image on a graphical display or as a balsa wood model, constructed section by section (Figures. 4-5).

RESULTS

HIV core

The specimens were examined by rotation and tilting in TEM to envisage that the major core structure was contained within the specimen. Thicker sections, between 60 nm and 80 nm, were chosen to enable analysis of the entire core.

A prevailing cone-shaped core was observed. Due to different section planes and angles different conical or round projections were generally observed in TEM (Figure 1a). With serially sectioned specimens round, centrosymmetric projections appeared with different diameter in consecutive sections and occasionally with a less electron dense central portion (Figure 1b). Goniometer analysis revealed that conical projections could also turn into bar-shaped projections in TEM by varying tilt and rotation angles.

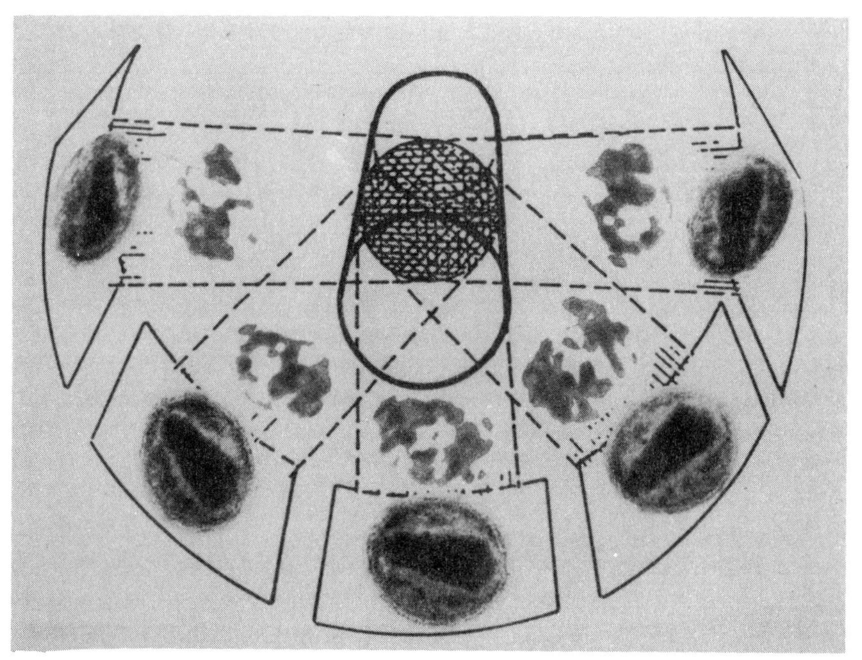

Figure 3.

Schematic schedule of the events of electronmicroscope tomography: Five different electronmicrographs of 60 nm thick specimens of an HIV particle, are obtained in TEM with tilt angles varying from -60° to plus 60°. Corresponding 2-dimensional density maps are obtained from digitized electron micrographs. Several consecutive discs will form a cylinder on which the reconstructed object is contained.

3-dimensional reconstruction

From 15 reconstructed, selected models the gross morphology of the HIV core showed a somewhat flattened, conical shape with the diameter of the broad end (60nm) about twice of that of the narrow end. Generally it was indicated that the structure was composed of two halves, divided by an open volume through the central portion of the core. There was also an indication of a somewhat skewed shape, structurally arranged along the long axis of the core model (Figures 4,5). By turning the core models their appearance was changed from conical into bar-shaped, and also a flattening of the structure at its narrow end was visualized (Figure 5).

In the reconstructed sections of both the internal part and a segment of the envelope (Figure 4) it was indicated that the core, apparently being situated in an open volume of the virion, was only connected to the surrounding, internal compartments by a structural joint between the narrow end of the core and the internal part of the viral envelope. There are indications that both structural halves of the core would be connected to a limited area of the viral matrix (Figure 4).

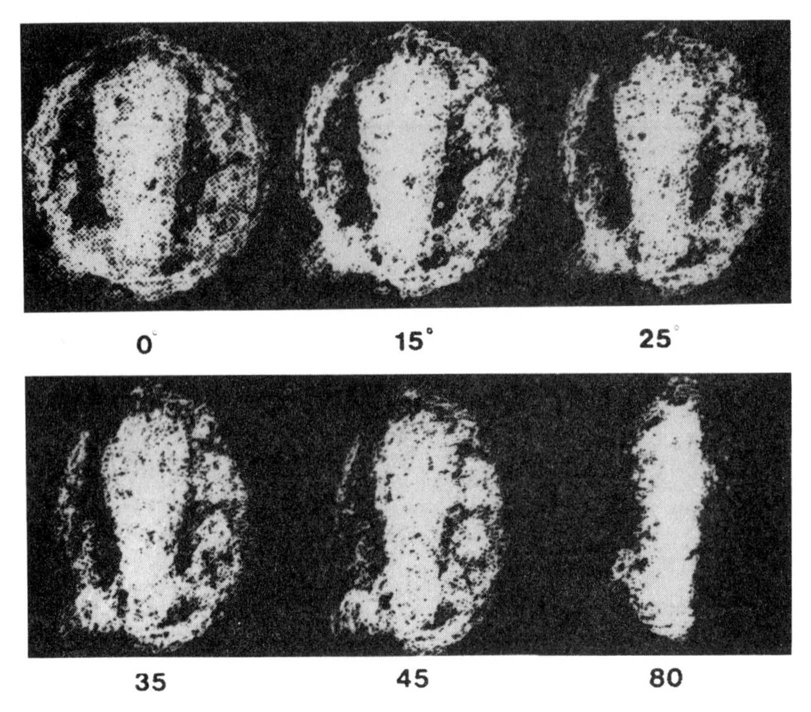

Figure 4.
Display images, taken at different **y** coordinates, of a reconstructed 60 nm section of an HIV particle. A big open volume is observed surrounding the core except for at its narrow end. The latter has a connection to the inner part of envelope. The content of the TEM section is visualized by rotation to 80°.

DISCUSSION

Projections in TEM of a population of sectioned HIV particles have shown that the majority of the cores of morphologically mature virions are cone-shaped (Figure 1a), in analogy with the cores of other lentiviruses (Frank *et al.*, 1978; Gelderblom *et al.*, 1987, 1989; Gonda, 1988; Palmer and Goldsmith, 1988; Grief *et al.* 1989). From this study image reconstructions also suggested a flattening of the narrow end of the core structure (Figure 5). It appeared with an irregular, rectangular shape, as observed from certain directions (Figure 5), which would accord with observations of bar-shaped projections in TEM (Gelderblom *et al.*, 1987). The observations in TEM of round projections of the core, differing in size with consecutive sections (Figure 1b), would reflect the views of tangential sections through the core.

It was elucidated from different views of the reconstructed models of the core that it would have a somewhat skewed structure containing two halves, however, no indication of a helical structure was attained.

Also cavities or holes were shown distributed in the entire structure (Figure 5) which would accord with the bright areas observed in TEM in the center of projections obtained from tangential sections of the core (Figure 1). In the reconstruction of a segment of the entire virion (Figure 4) there seems to be a big, open volume, most probably freed of a package of proteins, along the core.

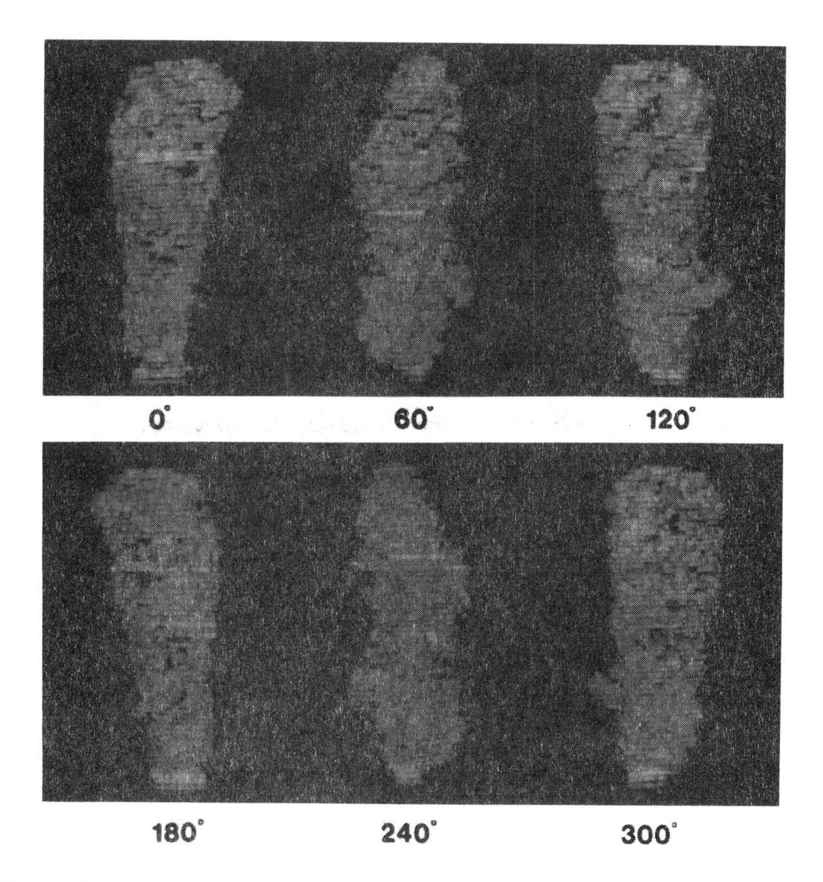

Figure 5.
A loose, somewhat skewed structure of another balsa wood model shows several cavities or holes possibly comprising two halves. The model is somewhat flattened at its narrow end (indicated by arrow).

The internal packaging of retrovirus has been judged by recombinant DNA manipulations (Delchambre *et al.*, 1989). The retro-secretion of recombinant HIV proteins (Wills, 1989) produced RNA-free particles (Lever *et al.*, 1989), possessing a structurally undefined assembly of internal proteins. Upon assembly of the core structure myristillation, i.e. incorporation of 14:0 fatty acid to MA, will occur (Veronese *et al.*, 1988).

In TEM MA has been observed as a 5 nm - 7 nm thick layer at the innermost part of the envelope of mature HIV (Gelderblom *et al.*, 1987). At early stages of virus maturation an assembly of structural/functional proteins of the core are observed inside the envelope in budding virions but also in released virus particles (Figure 1) (Gelderblom *et al.*, 1987; Gonda, 1988). On the reconstructed images of the sections through the entire, mature virus particle two halves of the narrow part of the core are indicated, connected to the inner part of the envelope by a relatively small structural joint (Figure 4). Most probably uncleaved p55 or the interaction between MA and CA (Veronese *et al.*, 1988) would be involved in such a possible joint of the core to the envelope. By this means the core, attached to the envelope, could protect its RNA genome and enzymes (Oroszlan *et al.* 1989; Haseltine, 1989) during the initial route of viral infection. The postulated connecting site between envelope and core could then be a target for future chemotherapy or subunit vaccination trials (Höglund *et al.*, 1989) to inhibit the production of mature, infectious virus.

REFERENCES

Chrystie, I.L., and Almeida, J.A. *AIDS*, **2**, 459-464, (1988).

Dahlberg, J.E. *Adv. Vet. Sci. Comp. Med.*, **32**, 1-35, (1988).

Delchambre,M., Ghysen,D., Thines,D., Thiriart,C., Jacobs,E., Verdin,E., Horth,M., Burny,A., and Bex,F. *EMBO J.*, **9**, 2653-2660, (1989).

Frank,H., Schwartz,H., Graf,T., and Schäfer,W. *Z.Naturforschung*, **33**, 124-138, (1978).

Gelderblom,H., Hausmann,E., Özel,M., Pauli,G., and Koch,M. *Virology*, **156**, 171-176, (1987).

Gelderblom,H., Özel,M., Pauli,G. *Archiv.Virol.*, **106**, 1-13 (1989).

Gonda,M.A. *J.Electron Micr. Techn.*, **8**, 17-40, (1988).

Grief,C., Hockley,D.J., Fromholc,C.E., and Kitchin,P.A. *J. Gen.Virol.*, **70**, 2215-2219, (1989).

Haseltine,W.A., Terwilliger,E.F., Rosen,C.A., Sodroski,G. in 'Retrovirus Biology. An emerging role in human diseases.' (R.C. Gallo and F. Wong-Staahl, eds.), M Dekker, Inc., (1988).

Haseltine,W.A. *J. AIDS*, **2**, 311-334, (1989).

Henderson,L.E., Copeland,T.D., Sowder,R.C., Schultz,A.M., and Oroszlan,S. in 'Human Retrovirus, Cancer and AIDS: Approaches to Prevention and Therapy', (D.Bolognesi, ed.), Alan R. Liss. Inc, New York, (1988).

Höglund,S., Öfverstedt,L.G., Gelderblom,H., Özel,M., Eriksson,M. *Proc. Ann. Meeting Scand. Neuropath. Soc. Iceland*, **13**, (1989).

Höglund,S., Dalsgaard,K., Lövgren,K., Sundquist,B., Osterhaus,A. and Morein,B. *Subcell. Biochem.*, **15**, 39-68, (1989).

Katoh,I., Yoshinaka,Y., Rein,A., Shibyua,M., Okada,T. and Oroszlan S. *Virology*, **145**, 280-292, (1985).

Leis,J. *et al.* J.Virol. **62**, 1808-1809, (1988).

Marx,P.A., Munn,R.J., and Joy,K.I. *Lab. Invest.*, **58**, 112-118, (1988).

Oroszlan S, and Luftig R.B. 'Retroviral Proteinases', Springer Verlag, Berlin-Heidelberg-New York, (1989)

Palmer,E, and Goldsmith,C.S. J. Electr. Micro. Tech. **8**, 3-15, (1988).

Roberts,M.M. and Oroszlan,S. *Biochem. Biophys. Res. Comm.*, **160**, 486-494, (1989).

Skoglund,U. and Danerholt,B. *Trends Biochem. Sci.*, **11**, 499-503, (1986).
Veronese,F.M., Copeland,T.D., Oroszlan,S,, Gallo,R.C., and Sarngadharan,M.G. *J. Virol.*, **62**, 795-801, (1988).
Wills,,J.W. *Nature*, **340**, 323-324, (1989).

16
Morphogenesis, Maturation and Fine Structure of Lentiviruses

Hans R. Gelderblom, Preston A. Marx, Muhsin Özel, Dirk Gheysen, Robert J. Munn, Kenneth I. Joy and Georg Pauli

Constituting one of the three subfamilies of the retroviruses the lentivirinae comprise genetically and structurally highly related agents (Frank, 1987; Gelderblom *et al.*, 1985a; Gonda *et al.*, 1985; Levy, 1988; Matthews, 1982; Weiss et. al., 1984). Lentiviruses are transmitted as exogenous agents, show a strong leucotropism and induce persistent, progressive infections *in vivo* (for review see Georgsson *et al.*, 1989; Haase, 1989; Levy, 1989; Narayan and Clements, 1989). Lentiviruses are cytopathogenic *in vitro* and partially serologically related. Primate specific lentiviruses share approximately 60% of homology in conserved proteins (Desrosiers *et al.*, 1989). Serological cross-reactivities localized mainly to the viral core, but also to the envelope, have been described for HIV and EIAV (Montagnier *et al.*, 1984; Montelaro *et al.*, 1988; Schneider *et al.*, 1986, 1987).

Lentivirus diseases of the ruminants, particularly the prototypic Maedi Visna of sheep, have long been confined to the interest of veterinary research. The emergence of the AIDS epidemic, however, and the rapid identification of the causative agent as a human specific lentivirus (Barre-Sinoussi et al., 1983; Gallo *et al.*, 1984; Levy *et al.*, 1984) has brought this virus subfamily into the focus of biomedical research.

Certain lentiviruses, like HIV and SIV, are able to infect specific subsets of T-lymphocytes in addition to cells of the mononuclear/macrophage lineage (Gendelman *et al.*, 1986, 1989; Rosenberg and Fauci, 1989). This dual host range might be an important factor for the induction of progressive immunosuppression which ultimately leads to the break down of the immune system and death of the infected individual.

The single stranded RNA genome of the human immunodeficiency virus (HIV) has been sequenced and its complex genomic organisation elucidated (Haseltine *et al.*, 1989). The genome of about 10.000 bases, like that of all other replication competent retroviruses, comprises three genes, coding for the structural proteins (*gag*, *pol*, *env*). The *gag* and env-coded proteins present in mature, cell released virions (Fig. 1; Leis et al., 1988) are processed from the respective $p55^{gag}$ and $gp160^{env}$ precursor proteins by endoproteolytic cleavage (for review see Oroszlan and Luftig, 1990; Putney and Montelaro, in press).

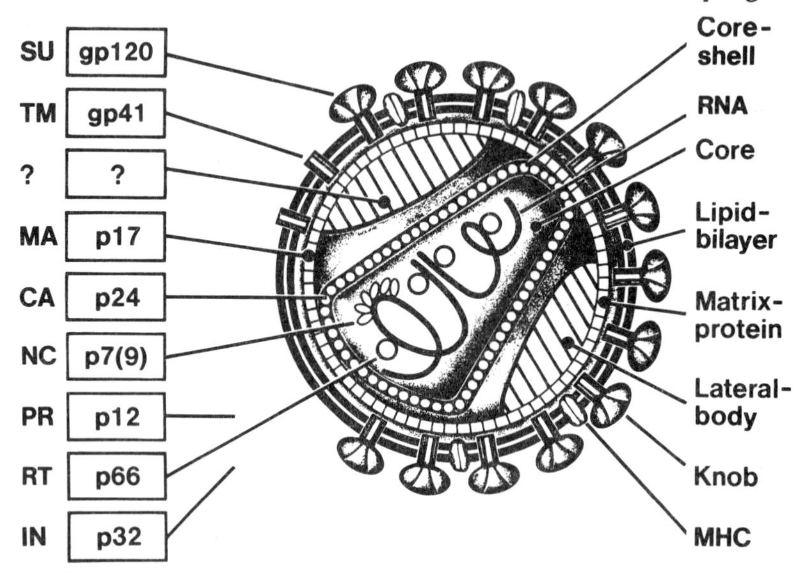

Figure 1.
Two-D model of HIV modified from Gelderblom *et al.* (1987b). Retrovirus structural proteins have been designated by the letter 'p' for protein or 'gp' for glycoprotein followed by the molecular weight of the protein (x10-3) (August *et al.*, 1974). Recently a more general nomenclature for the structural proteins was proposed (Leis *et al.*, 1988) which uses a two-letter acronymic code describing either the location of the protein in the virion, the enzymatic activity and/or biological function. MA stands for matrix protein, CA for capsid, NC for nucleocapsid, SU for surface, and TM for transmembrane protein, respectively. PR denotes the viral protease, RT reverse transcriptase, and IN the viral integrase. This code is comparable to that of other enveloped viruses and helps to denominate respective proteins in the different retrovirus strains or subfamilies.

On the genome of the primate immunodeficiency viruses until now six additional genes, obviously involved in virus regulation, were identified. The proteins coded by these genes regulate the virus cell-interaction at different levels during the infection cycle in a higly intricate way. Two of them, *tat* and *rev* are essential for efficient virus production, whereas *nef*, *vif*, *rap* and *out* seem not to be required for HIV-1 replication *in vitro* (Cullen and Green, 1989; Dahlberg, 1988; Gallo *et al.*, 1988; Haseltine *et al.*, 1988, 1989).

Although the biochemical, structural and morphological components of lentiviruses, particularly of HIV, have been characterized in great detail (Bouil lant *et al.*, 1984; Gelderblom *et al.*, 1985a, 1989; Haseltine et al., 1989; Parekh *et al.*, 1980; Putney and Montelaro, 1989) little is known about the biochemical processes necessary for the different steps in retrovirus assembly and maturation. The aim of this article is to present structural details of lentiviruses, with special emphasis on HIV, and to correlate these morphological observations with possible functional concepts, which are derived also from biochemical studies. It is hoped that from such unifying approaches new ideas for immunological and chemotherapeutic intervention strategies can be deduced.

MORPHOLOGY OF LENTIVIRUSES

A number of recent reviews have dealt with questions of comparative lentivirus morphology (Bouillant and Becker, 1984; Frank, 1987; Gelderblom *et al.*, 1 985a, 1988, 1989; Gonda, 1988; Gonda *et al.*, 1987; Hockley et al., 1988; Ladhoff *et al.*, 1986; Lecatsas *et al.*, 1984; Munn *et al.*, 1985; Palmer and Goldsmith, 1988). Morphological criteria have been used already a decade before the description of the reverse transcriptase to distinguish between different RNA-tumor virus isolates (Bernhard, 1960). This morphological classification (Figure 2) parallels biological differences and was therefore extended whenever new retrovirus isolates were described (Harven, 1974; Matthews, 1982; for review see Gelderblom *et al.*, 1989). Distinctive criteria for class ification, assessable mainly by thin section transmission electron microscopy (TEM), are enlisted in Table 1.

Table 1.
Structural criteria for the morphological classification of retroviruses (modified from Gelderblom et al., 1985a)

Morphogenesis		
	A	Cores preformed in the cytoplasm
	B	Core assembly concomitant with budding
	C	Distance of core shell to prospective envelope
	D	Dimension of the RNP core complex
Morphology		
	E	Shape of the core (isometric or tubular)
	F	Site of the isometric core (con or eccentric)
	G	Dimension of envelope projections
	H	Anchorage stability of envelope projections

The current knowledge about the fine structure and the antigenic composition of morphologicaly mature HIV are summarized in Figures 1 and 3.

Mature lentiviruses, like HIV, SIV, EIAV etc. are enveloped by a lipid bilayer derived from the host cell (Aloia *et al.*, 1988), spherical or slight ly ovoid in outline with a diameter of 130 to 160 nm. The viral surface is studded with knob-like projections 9 to 10 nm in hight and 14 to 15 nm at the outer circumference. These surface knobs are particularly well demonstrated by tannic acid treatment (Figs. 3, 4) of the specimen before embedding followed by conventional heavy metal contrasting (Gelderblom *et al.*, 1987b, 1989; Schidlovsky *et al.*, 1978). However, in the high contrast introduced by the tannic acid, the two distinct electron dense lines limiting the RNP shell visible under conventional preparation conditions are lost (Figures 3 - 5) .

The knobs consist out of the *env* coded gp120 SU protein which are anchored only loosely to the virion. The gp41 TM protein, likewise env coded, and host cell derived proteins are firmly inserted into the bilayer.

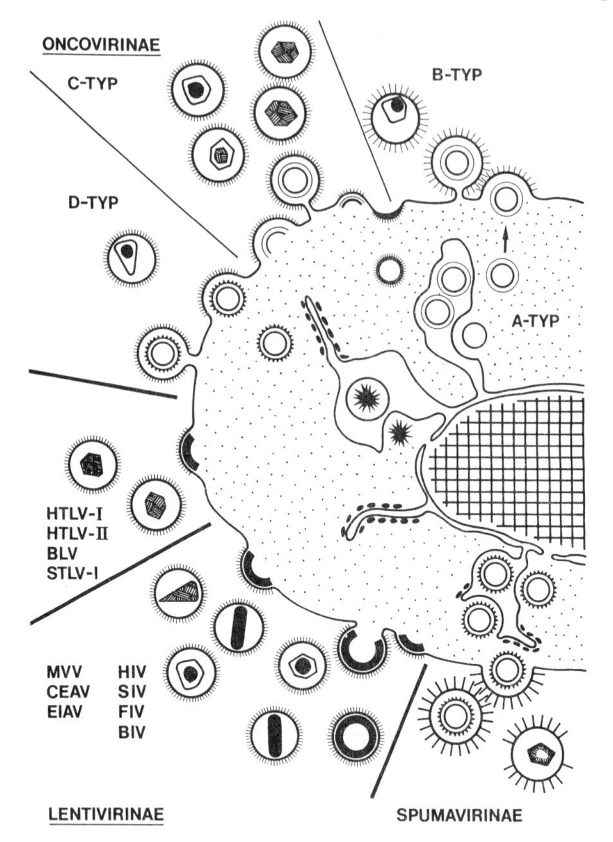

Figure 2.
Schematic presentation of the retrovirus family (modified from Gelderblom *et al.*, 1985a). Particles of types A to D comprise the oncovirus family of the retroviruses (Bernhard, 1960; Harven, 1974). A-type particles are morphologically preformed cores of the mouse mammary tumor, of the prototype B-type particle (Sarkar, 1987). These cores migrate to the plasma membrane and during budding become enveloped to form the complete, doughnut shaped, immature virion. This, after morphological maturation, contains an eccentrically located, condensed isometric core and is studded with surface projections about 10 nm in length. C-type viruses cause sarcomas and lymphomas in chicken, cats and mice. Their core is assembled parallel to the budding process. The mature C-type virion is studded with surface projections about 5 nm in length and contains a concentrically located isometric core. D-type particles are found in certain primates and can also be related to malignancies (Fine and Schochetman, 1978). Virions bud by the envelopment of preformed cores and show surface projections 5 nm in length. Mature particles are characterized by elongated cores. HTLV-I and related viruses represent an inter mediate group of retroviruses with morphological features common to both C-type and lentiviruses. The immunodeficiency viruses of man (HIV), monkeys (SIV), feline, and bovine origin have their counterparts in the lentiviruses of the ruminants, e.g. Maedi Visna (sheep), caprine

encephalitis arthritis virus, and of horses (equine infectious anemia virus). They bud concomitant with the assembly of the viral core and show surface knobs 9 to 10 nm in length. The electron dense core shell of immature lentivirus is condensed in mature particles to a cone shaped core. The pathogenic potential of the spumavirinae is unknown (Gelderblom and Frank, 1987; Weiss, 1988). They occur frequently as contaminants in cell cultures as foamy viruses, derived from organs, such as kidneys. Spumaviruses show envelopment of preformed cores, which after release of virions never fully condense and are studded with surface projections about 10 nm in length.

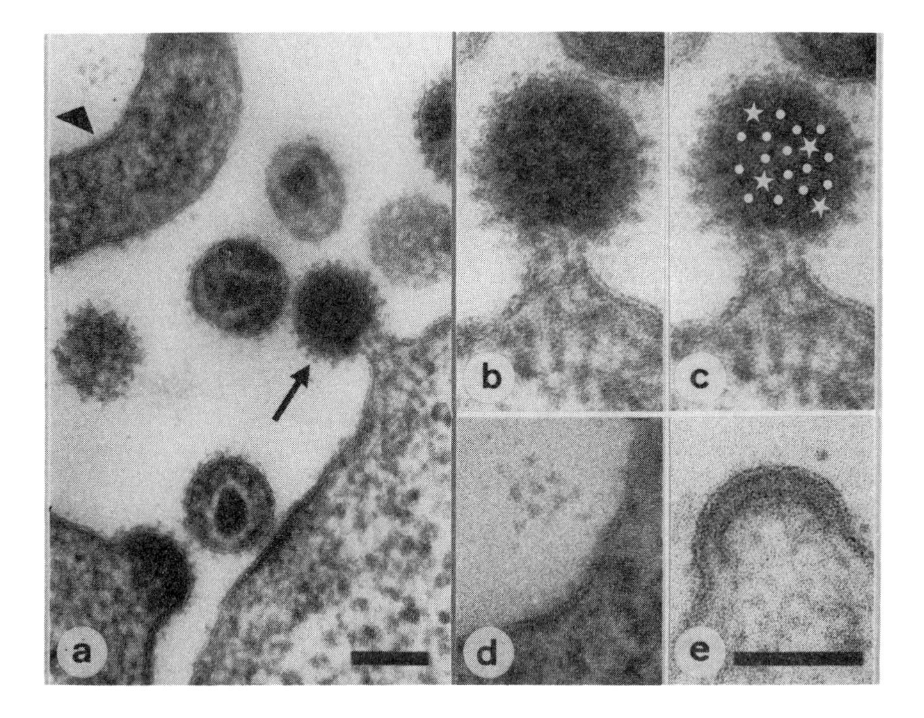

Figure 3.
Morphogenesis and surface architecture of HIV. Thin section TEM was performed with (a–d) or without (e) tannic acid treatment. The budding virion (ar row) was analyzed by image rotation enhancement, and shows a T=7 *lævo* symmetry of the SU knobs. (b, c) were reversed for correct handedness, which was deduced from surface replica specimens (Özel *et al.*, 1988). SU knobs, when cut tangentially (triangle in a) show at higher magnification a trimeric outline. (e) without tannic acid pretreatment, the envelope knobs are only scarcely detectable, while the structure of the underlying *gag* shell becomes clearly visible (from Gelderblom *et al.*, 1989). Bars =100 nm.

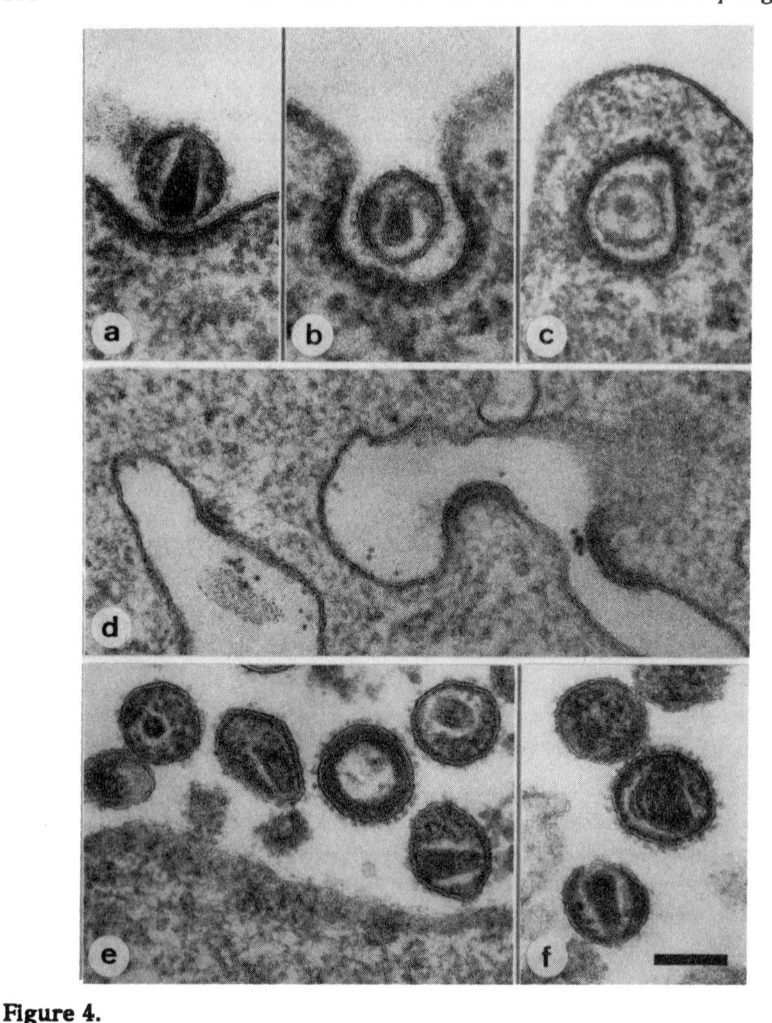

Figure 4.
Thin section TEM illustrating steps in the life cycle of HIV-1 and HIV-2. (a - c) demonstrating consecutive steps in receptor mediated uptake show interaction of knobs and the surface of coated pits/vesicles. (d) early budding HIV-1 prepared without tannic acid pretreatment. (e) and (f) reveal different steps in maturation. In (e) an immature HIV-2 particle with a concentric core shell and a dense fringe of SU knobs is seen. An intermediate step in core condensation is shown in (f) for HIV-1, which by morphological means is indistinguishable from HIV-2. Bar = 100 nm

Directly apposed to its inner leaflet the *gag* coded p17 MA protein is forming a shell 5 to 7 nm in thickness. Depending on section plane and angle, the cores of lentiviruses appear cone-shaped, tubular or centrosymmetric in TEM. When full in section plane, the cone-shaped core is spanning the entire diameter of the virion. It measures 55 to 62 nm at the broad end, which is always surrounded by an electron-lucent space, and 25 nm at the narrow end, which often shows continouity with the MA shell. Being 120 - 130 nm in lenght the core is often bulging the viral envelope to an ovoid outline. Besides this predominant cone-shaped structure, a small percentage of tubular cores with a slightly smaller diameter of 40 nm can be seen. These tubular cores are usually longer, up to 200 nm, than the cone-shaped cores. Occasionally several cores can be seen within one virus particle. The capsid of the core is composed of p24 CA. The architecture of the lentivirus core was analysed recently by EM tomography (Höglund *et al.*, this volume) as a slightly flattened, wedge-like cone showing binary symmetry.

Two molecules of genomic RNA together with the *gag* proteins p7 and p9 (Veronese *et al.*, 1988) are forming the RNP enclosed by the core capsid. Parallel to the long axes of the core so called lateral bodies are situated, electron dense masses of unknown composition and function. In centrosymmetric, end-on views of the virion often a penta- or hexagonal 'window' can be seen around the core (Marx *et al.*, 1988).

MORPHOGENESIS OF LENTVIRUSES

Less is known about the detailed interaction of the structural components of these agents during assembly and morphogenesis which takes place in a budd ing process at cellular membranes. As the first step, an electron dense shell 25 to 30 nm in thickness can be seen directly underneath the lipid bilayer of the plasma membrane (Figures 3, 4). This shell, covered by the plasma membrane, is growing to a complete sphere about 130 to 145 nm in diameter, enclosing also cytoplasmic constituents. Without tannic acid pretreatment the shell shows two well defined concentric dense rims 15 to 18 nm apart, the outer one directly apposed to the lipid bilayer. From the stain distribution one could assume that the outer rim is representing the myristylated N-terminus of the $p55^{gag}$ precursor, while the inner one reflects the two molecules of genomic RNA associated with the C-terminus of p55. Using the formula $^4/_3\pi r^3$ and the dimensions given above, a volume of 127,000 to 524,000 nm^3 can be calculated for the spherical p55 shell. This high variation does not encourage to speculate further on the number of p55 molecules present in that shell.

However, it does not seem unlikely that the 'lateral bodies' present in morphologically mature HIV, SIV and EIAV represent surplus structural proteins not used for the formation of the matrix shell, the capsid and the RNP complex. Alternatively, these dense bodies might constitute virus coded regulatory proteins, like *vpx* (Henderson *et al.*, 1988). Before a final model for the capsid architecture of lentiviruses can be established, obviously more reliable stoichiometric data have to be collected.

Concomitant with the assembly of this internal ribonucleoprotein complex (RNP), projections on the surface of the bud become visible (Figures 3, 4). After completion of the core shell, the enveloped immature particle with a diameter of 150 to 160 nm is released from the cell showing still its morphologi cally immature organization (Figures 4), e.g.

the concentric *gag*-RNP-shell (Bouillant and Becker, 1984; Filice *et al.*, 1987; Gelderblom *et al.*, 1985a, 1988; Hockley *et al.*, 1988; Katsumoto *et al.*, 1987; Nakai *et al.*, 1989; Palmer and Goldsmith, 1988).

Later, probably in the course of minutes to hours, the inner constituents rearrange to the condensed core of the morphologically mature lentivirus. A model for the processing of the *env* and *gag* polyproteins and their rearrangement was proposed by Bolognesi *et al.*,(1978). The morphological transition is fast, as intermediate steps are observed rarely (Figure 4f) (Gelderblom *et al.*, 1985a, 1989). Morphological maturation is accompagnied by a decrease in viral diameter to 130 to 160 nm (Gelderblom *et al.*, 1985a, 1989; Bouillant and Becker, 1984; Munn *et al.*, 1985).

Although a detailed description of the morphogenesis and fine structure of retroviruses in general and especially of the lentivirus subfamily has been achieved there are still essential questions to be answered:

1. What does the infectious virion look like ?
2. How many SU knobs have to be present to render a particle infectious ?
3. Is the morphologically mature virion infectious ?
4. When does the processing of the $p55^{gag}$ protein takes place?
5. What are the mechanisms involved in the morphological maturation, leading to the transition of the sperical shell to the cone shaped core and to the icosahedral MA protein layer?
6. What is the molecular nature of the 'lateral bodies' ?

The application of recombinant DNA technology has already allowed to elucidate some essential steps in virus assembly and maturation (Figures 5, 6). HIV and SIV *gag* constructs, when inserted into the baculovirus vector and propagated in the insect *Spodoptera frugiperda* host cell system showed expression and selfassembly of the precursor *gag* protein into virus-like particles at the cell membrane despite the absence of viral RNA, viral enzyme and envelope protein (Delchambre *et al.*, 1989; Gheysen *et al.*, 1989; Jacobs *et al.*, 1989; Overton *et al.*, 1989). The core shell of these particles resembles closely that of immature lentiviruses. They do not undergo morphological maturation, hence they contain uncleaved $p55^{gag}$ molecules which in the budding process obviously have to interact with their myristilated amino terminus with the lipid membrane of the host cell. The result of this interaction is the formation of enveloped, immature particles which are released from the cell to a certain degree. They measure 135 to 150 nm in diameter and can be isolated for further characterization.

Proteolytic cleavage of the *gag* precursor apparently is one prerequisite for the rearrangement of the inner structural constituents to the core of mature lentiviruses (for review see Oroszlan and Luftig, 1990). Due to the lack of the viral proteinase in this construct, the p55 particles cannot become processed further. The absence of viral nucleic acid, however, might make such particles an interesting immunogen.

Another prerequisite for retrovirus formation is the myristylation of the $p55^{gag}$ precursor protein (Rein *et al.*, 1986; Oroszlan and Luftig, 1990). After deletion of the myristylation site at the N-terminus of the *gag* precusor, this modified p55 polyprotein is not able to associate with cellular membranes. Instead, self-assembly to spherical particles, reminiscent of A-type particles (Figure 2), takes place in the cytoplasm (Figure 5a). Thus, myristylation appears essential for correct targeting of the p55 polyprotein.

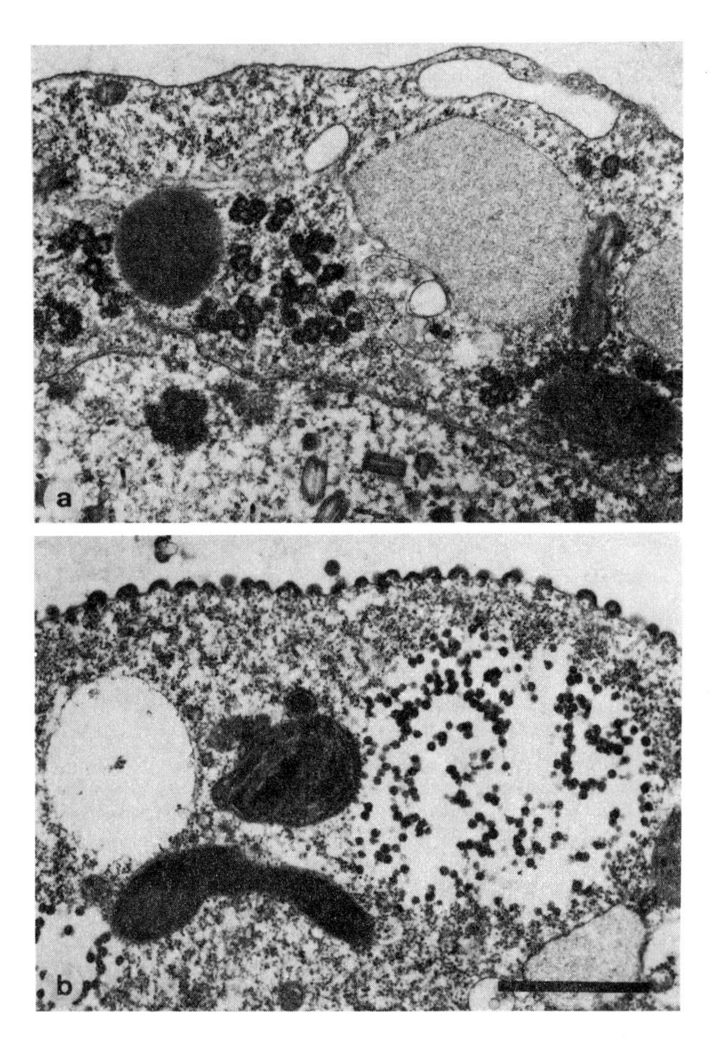

Figure 5.

Spodoptera f. cells transfected with different baculovirus-HIV *gag* constructs. (a) after deletion of the myristylation site at the *gag* protein, spherical structures, reminiscent of immature retrovirus cores, are formed in the cytoplasma as well in the nucleus without any apparent affinity to membranes. (b) in contrast, the myristylated *gag* precursor protein assembles into budding structures at the plasma membrane. Bar = 1 μm

Heterologous gene expression in *Spodoptera* cells of a truncated *gag* construct containing only the p17-24 precursor gives rise to aberrant 'budding' of long tubular membrane bound structures. This implies that the deletion of the C-terminus (p15) of the *gag* precursor protein abolishes 'spherical' particle formation, although membrane targeting of the p17-24 protein and membrane evagination still occurs (Gheysen *et al.*, 1989). Furthermore, site directed mutations into the p17-24 cleavage site of the precursor protein alter virion morphogenesis giving rise to 'oversized' capsid structures and prevent virus replication (Göttlinger *et al.*, 1989). Finally, for correct packaging of the viral RNA a region close to the translation start of the *gag* protein is necessary (Lever *et al.*, 1989).

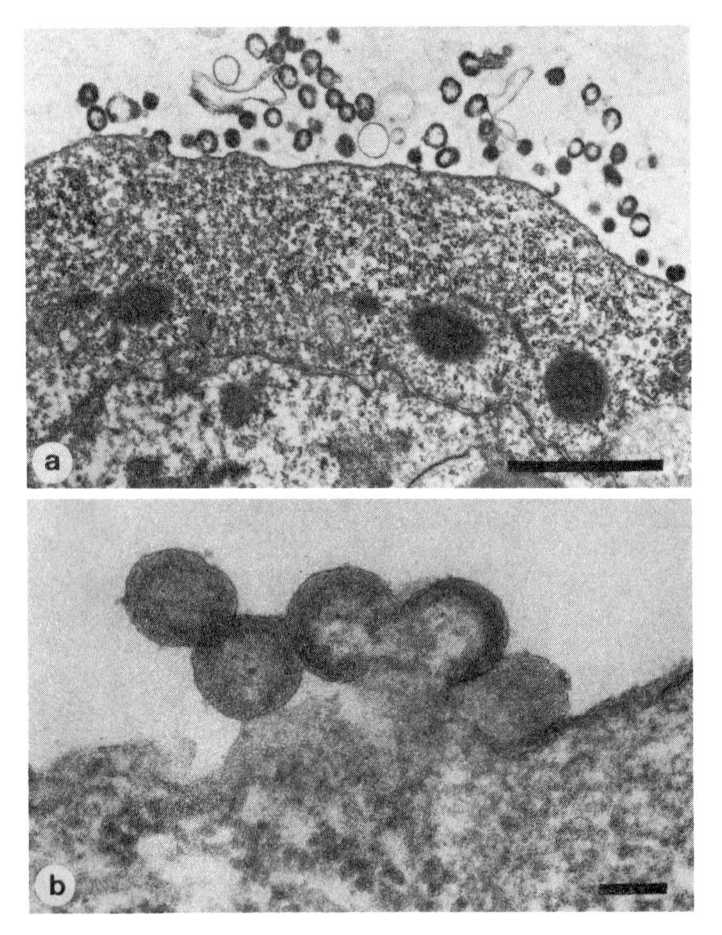

Figure 6.
Release of p55 particles from *Spodoptera f.* cells transfected with a complete HIV-1 *gag*-baculovirus construct (a). The particles show at higher magnif ication (b) an immature, spherical core shell, but no evidence for surface knobs. Bars = 1 μm (a) and = 100 nm (b).

MORPHOLOGICAL MATURATION

It is not known when the virus coded protease is active, and it might be speculated that the protease is directly involved in the morphological transition of the immature to the mature core. However, it is also well conceivable, that the immature particles contains already the cleaved polypeptides of the mature virion, however, still in their original position.

Under specific preparation conditions, in case of Friend murine C-type oncoviruses, immature cores have been observed regularly, and were therefore considered as the preserved so called 'native' core (Frank *et al.*, 1978). If their findings can be generalized, we would have to assume that the condensed core structures observed in thin section EM represent fixation artifacts. On the other hand, if the processing of the *gag* proteins has occurred already before viral release, rearrangement of the processed proteins, leading to the condensation of the NC and RNP, would need additional factors, e.g. a change in pH or osmolarity. On the other hand, as cone-shaped cores have been demonstrated also by negative staining EM in case of HIV, SIV, and EIAV, this type of symmetry unknown hitherto in virology does not represent a preparative artifact (Chrystie and Almeida, 1989; Gelderblom *et al.*, 1988; Grief *et al.*, 1989; Hockley *et al.*, 1988; Palmer *et al.*, 1988; Roberts and Oroszlan, 1989).

ENVELOPE CONSTITUENTS: VIRUS AGEING

The knobs on the envelopes of lentiviruses measure 9 to 10 nm in height and have a broad, slightly convex surface 14 to 15 nm in diameter (Gelderblom et al., 1987b; Grief *et al.*, 1989; Hockley *et al.*, 1988; Palmer *et al.*, 1988). SU and TM of HIV are processed by endoproteolytic cleavage from a heavily glycosylated gp160 precursor (Haseltine *et al.*, 1989; McCune *et al.*, 1988) using a cellular protease.

Compared to other enveloped viruses, particularly of the oncovirus subfamily of the retroviruses, lentivirus SU knobs are only weakly anchored to the viral surface resulting in their easy loss (Gelderblom *et al*, 1985a; Schneider *et al.*, 1986; Weiland and Bruns, 1980). Whereas SU of oncoviruses are covalently linked by disulfide bonds to TM (Pauli *et al.*, 1978; Pinter and Fleissner, 1977; Schneider and Hunsmann, 1978), SU of HIV is bound to the virion only by non-covalent, i.e. weak bonds (Modrow *et al.*, 1987). This weak anchorage obviously causes the extensive, spontaneous loss of the surface proteins of HIV, SIV and MVV observed in cell cultures. This loss might also occur *in vivo*, where released SU ultimately might induce some of the immunopathogenic events observed during the course of the disease (Clayton *et al.*, 1989; Gelderblom *et al.*, 1985b, 1987a, 1988; Lanzaveccia *et al.*, 1988; Schneider *et al.*, 1986; Rosenberg and Fauci, 1989; Siliciano *et al.*, 1988).

VIRUS ENVELOPE: INTERACTION BETWEEN SU, TM, AND MA ?

Using optical rotational enhancement (Markham *et al.*, 1963) for the analysis of the surface architecture of HIV and SIV a number of 72 knobs per virion was calculated, arranged in a T=7 *lævo* symmetry (Gelderblom *et al.*, 1987b; Özel *et al.*, 1988; Takahashi *et al.*, 1989). This number of knobs is corroborated by direct counts using viral surface replicas (Özel *et al.*, 1988). It should be mentioned that, due to the inherent limitations of the enhancement technique, this symmetry is controversally discussed.

There is ample evidence accumulating for an oligomeric structure of the lentivirus TM and SU proteins. The dimensions of the knobs in TEM are too large to account for a single SU gp120 polypeptide as the morphological entity. From the calculated number of 72 surface projections per virion and the stoichiometric analysis of polypeptides of EIAV (Parekh *et al.*, 1980) a number of 3 to 4 polypeptides per morphological unit can be deduced. Direct infor mation on the subunit structure is coming from TEM thin section and negatively stain TEM. Triangular structures can be observed on the surface of HIV and SIV (Gelderblom *et al.*, 1988, 1989; Grief *et al.*, 1989).

Finally, and more convincing for molecular biologists, evidence for an oligomeric structure was obtained by chemical cross-linking experiments pointing to a tri- or tetrameric assembly of the SU and TM protein of SIV and HIV (C. Grief, pers. communication; Pinter *et al.*, 1989; Schawaller *et al.*, 1989) .

The regular arrangement of surface structures might be induced by the interaction of TM with MA, which in morphologically mature particles forms an iso metric scaffold underneath the viral lipid bilayer (Marx *et al.*, 1988). In previous studies we have used a computer graphics technique that emulates thin section electron microscopy to test various morphologic models of HIV. The general scheme for constructing and testing models is presented in Table 2. The specific parts of the virion model were approximated with polygons. Each model was randomly distributed in space and 'clipped' to simulate thin sections (Foley and van Dam, 1982). The diameter of each computer-generated 'thin-section' was clipped to 40% of the diameter of a retrovirus model. The models were rendered using a scan line hidden-surface algorithm both for wire frame models (Fig. 7) and shaded solid models (Figs. 8 to 10). The com puter generated images were produced on a VAX-8600, a Ridge 32, and a Sun Microsystems 2/120. The shaded versions of the images were initially rendered on a Raster Technologies 1/20 graphics frame buffer, and photographed through a Matrix PCR-35 Camera. In earlier studies (Marx *et al.*, 1988) this technique was used to construct and test models of HIV. Conical cores enclosed by various types of regular polyhedrons were prepared. Arrays of these computer-generated models of possible HIV structures were randomly oriented and 'thin-sectioned' using the three-dimensional clipping procedure. About 2000 'thin-sections' of each proposed model were compared to electron micrographs of about 2000 virions of HIV. By a process of trial and error we were able to 'rule in' or 'rule out' a specific polyhedron as well as the specific orientation of the core inside the polyhedron. Figure 7 illustrates the process. This technique of computer emulation of thin-section electron microscopy predicted that (a) the virion structure of HIV consisted of an envelope enclosing an icosadeltahedral capsid and, (b) the rod-like core is placed inside the icosadeltahedron so that it abuts 2 opposing hexagonal vertices.

Table 2.
Steps for testing virus models by computer emulation of electron microscopy

1.	A specific 3 D model of a retrovirus is constructed from polygons and entered into a computer graphics program.
2.	The model is randomly positioned in 3 dimensional space
3.	The model is sectioned to desired thickness by 3 dimensional clipping e.g. 40% of model thickness.
4.	The process is repeated to produce many hundreds or thousands of views of the sectioned model.
5.	The model is tested by comparison to electron micrographs.

As an example how this technique can be used, in Figures 8 and 9, a set of 3 criteria were applied to a basic assumption about the structure of type B retroviruses. The hypothesis that the type B virus core is eccentrically located was tested. Figures 8 and 9 illustrate two possible positions (random versus eccentric) for the inner core of type B virus. The inner core is randomly distributed in Figure 8 and is eccentrically placed in Figure 9. In Figure 8, the inner core is visible in 31 of 64 particles. Of these 31, the core is eccentric in only 4 of the models or 12,9%. In Figure 9 the core is placed in an eccentric location. In the eccentric model 33 of 64 virions display the core. The core is eccentric in 16 or 48,4 %. By actual observation in the electron microscope the number of particles with eccentric cores ranged from 50 to 57 %. The computer model therefore predicts that the core is eccentrically placed in type B retroviruses. Table 3 presents a summary of the criteria that were used to test this simple hypothesis. By trial and error this technique is useful for building virion models that can be tested.

Figure 10 shows a further improvement of computer model testing. In this procedure the virion model is sequentially sectioned. The upper panel shows the first section. The corresponding squares in the second and third panels show the next two sections of the same virion. This allows comparison of virions or cell organelles to sequential sections. This is the most powerful of the techniques, because correct matches can prove that a particular model is accurate.

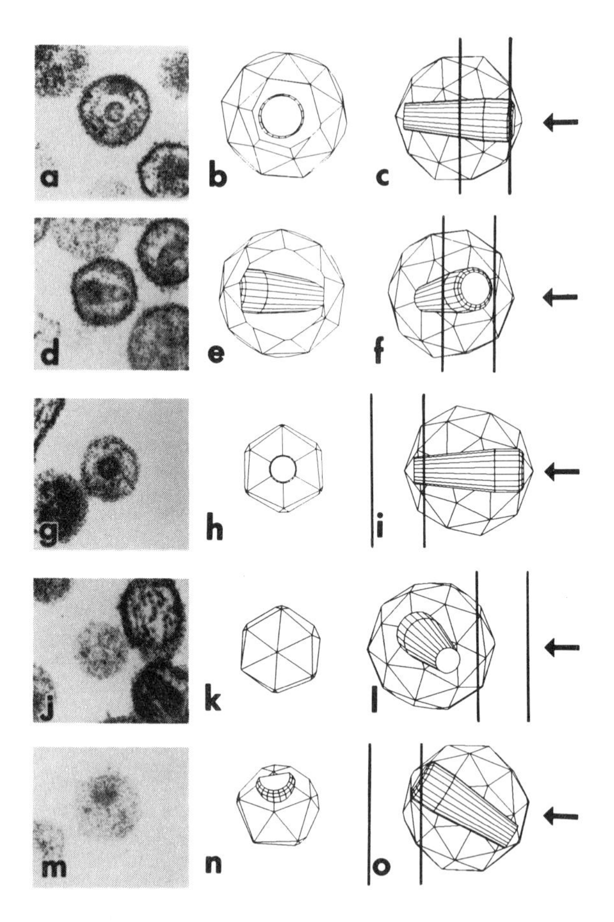

Figure 7.
Electron micrographs of HIV (a, d, g, j, m) are placed next to matching computer-sectioned envelope-associated capsids (b, e, h, n). To help visualize how these structures could result from thin-sectioning, the bold lines in c, f, i, l, and o show the thin-section cut through the virion. The arrows show the viewing direction. All parts of the envelope-associated capsid were approximated using polygonal models. Three-dimensional clipping was used to perform the computer sectioning. The models were rendered using a scan line hidden-surface algorithm. The computer images were produced on a Sun microsystems work station and printed on a Image laser printer.

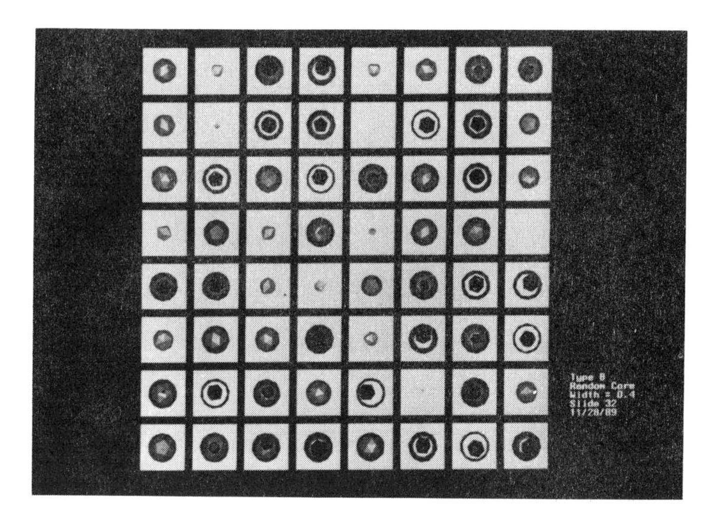

Figure 8.
A hypothetical model of the core of type B oncovirus. The model hypothesizes that the virion is composed of an outer envelope, an envelope-associated capsid and an inner core. The computer emulation of thin electron microscopy shown here test the location of the inner core in 64 examples of the model. In this model an icosahedral core was randomly placed inside an icosadeltahedral capsid. The surface is not shown.

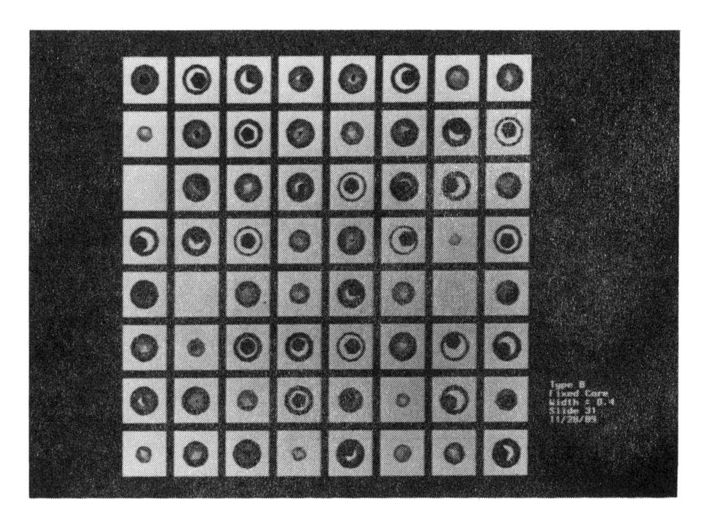

Figure 9.
The same model as in Figure 8, except that the inner core is fixed eccentrically instead of being randomly placed.

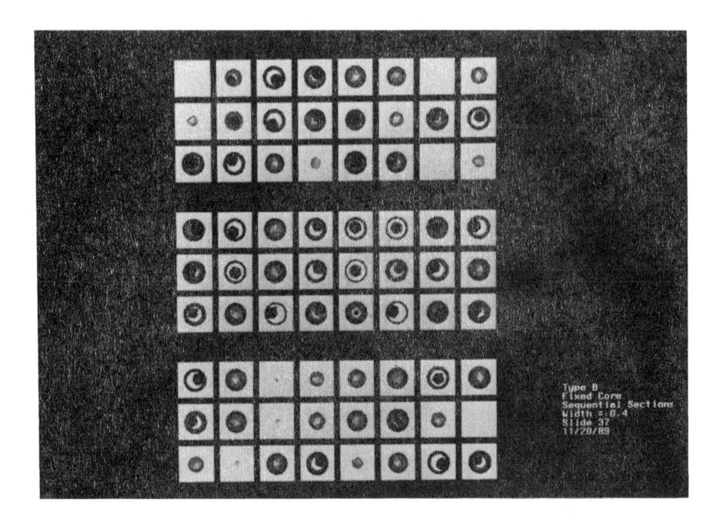

Figure 10.
Twenty-four examples of the model in Fig. 8 are shown sequentially sectioned. The top panel shows the first 'thin-section' through an array of 24 mode ls of the type B retrovirus. The second panel shows the next section of the same array and the third section is shown in the bottom panel. Each section is 40% of the diameter of the virion model. The figure is read by viewing the corresponding squares in the upper, middle, and lower panel. If the panel is empty, it means that the virion model fell outside the section plane.

Table 3.
Three criteria for testing computer generated models of retroviruses.

Criteria	Examples
Testing frequency of core position	Type B retrovirus - What percentage have an eccentric core ? What percentage has a central core? Do the frequencies agree with those observed in EM ?
Correlation of window shapes and their frequency	Do window shapes and their frequency correlate precisely with EM ? Percent hexagonal, etc.?
Shape of the inner core	Do core fragments agree with EM?

VIRUS ENVELOPE: HOST COMPONENTS

Biochemical and immunological analysis of purified virions as well as immuno-EM show that host cell derived proteins are integral parts of the viral en velope (Gelderblom *et al.*, 1987a; Henderson *et al.*, 1987; Hoxie *et al.*, 1987; Kannagi *et al.*, 1987). Of special interest is the stable insertion of MHC class I, and especially of class II, determinants in the viral bilayer. These proteins are involved in the regulation of the immune system. The interaction of MHC class II protein (HLA-DR) with the CD4 receptor is essential for the activation of CD4 positive lymphocytes being the main target of HIV. Latently infected CD4 cells can become immune activated resulting finally in the productive, cytocidal infection. Furthermore CD4 serves as the cellular receptor for HIV on T cells and macrophages (Dalgleish *et al.*, 1984; Gendelman *et al.*, 1989; Klatzmann *et al.*, 1984). The determinants involved in binding of HIV-SU have been confined to the V1 domain of CD4 (Arthos *et al.*, 1989). Binding of MHC class II proteins covers a larger region of the CD4 molecule comprised of domains I and II (Clayton *et al.*, 1989). From these results a direct interference of SU and MHC-proteins appears unlikely.

However, it is not unreasonable to assume, that the MHC antigens on the viral envelope might help the virion to act as a 'wolf in a sheeps clothing', and particularly, after shedding of the SU knobs, to serve as a ligand to the CD4 receptor. In addition, CD4 might interfere with MHC-proteins expressed on antigen presenting cells thus leading to a functional impairment of the immune system.

VIRUS ENTRY

The host range of HIV and SIV apparently is determined by the strong interaction of SU with the cellular CD4 receptor (Arthos *et al.*, 1989; Lasky *et al* ., 1987). This interaction obviously is the essential step in initiating virus entry. Whether other cellular surface proteins, like e.g. the Fc or the complement receptor in connection with antibodies directed against virus surface proteins are involved in virus entry is still in question (Bolognesi, 1989; Sölder *et al.*, 1989).

Entry *in vitro* takes place by either one of two mechanisms: Direct fusion with the plasma membrane (Marsh and Dalgleish, 1987; Marsh and Helenius, 1989; McClure *et al.*, 1988; Ribas, 1988; Stein *et al.*, 1987) or gp120-CD4-receptor-mediated uptake (Figure 4 a-c) which is assumed to be followed by fusion within the endosomal compartment (Pauza *et al.*, 1988). Both mechanisms have been observed to occur independently from each other in one and the same cell (Goto *et al.*, 1988; Grewe *et al.*, 1990). The different entry modes might be explained by the varied amount of SU protein present on the viral envelope. When enriched virus suspensions are used in entry studies, mainly fusion can be observed (Grewe *et al.*, 1990), probably because most of the kno bs are lost during virus purification. These 'naked' virions with the exposed hydrophobic portions of their TM proteins should be able to interact more readily with the plasma membrane of the host cell resulting in direct fusion. The domain responsible for membrane fusion was identified as a highly hydrophobic region near to the amino-terminus of TM gp41 (Bosch *et al.*, 1989). The virion studded with a higher number of SU knobs will gain entry by rec eptor mediated uptake (Figure 4 a - c). At present, the question which of the observed entry mechanisms leads to effective infection is unresolved. It might be answered in part by the final characterization of the infectious virus particle. Likewise, ways necessary for pharmaco-therapeutic intervention might become paved after detailed stoichiometric analysis of the virion as well as elucidation of the events during particle assembly and maturation.

ACKNOWLEDGEMENTS

The authors thank Bärbel Jungnickl and Bernd Wagner for the careful preparation of the photografic prints, Wolfgang Lorenz for drawing the schemes of Figures 1 and 2, and Regina Scheidler, Monika Ewald-Feld, Hilmar Reupke and F. de Foresta for their dedicated technical assistance. Part of the work pres ented was supported by a grant of the Ministry of Research and Technology, Bonn to H.G., and a grant of the Belgian IERSIA-IWONL to dg. The authors are grateful to Dr. M. de Wilde for supporting part of the research in his department.

REFERENCES

Aloia, R.C., Jensen, F.C., Curtain, C.C., Mobley, P.W. and Gordon, L.M. *Proc. Natl. Acad. Sci. USA*, **85**, 900-904 (1988).

Arthos, J., Deen, K.C., Chaikin, M.A., Fornwald, J.A., Sathe, G., Sattentau, Q.J., Clapham, P.R., Weiss, R.A., McDougal, J.S., Pietropaolo, C., Axel, R., Truneh, A., Maddon, P.J. and Sweet, R.W. *Cell*, **57**, 469-481 (1989).

August, J.T., Bolognesi, D., Fleissner, E., Gilden, R.V. and Nowinski, R.C. *Virology*, **60**, 595-601 (1974).

Barré-Sinoussi, F., Chermann, J.C., Rey, F., Nugeyre, M.., Chamaret, S., Gruest, C., Daquet, C., Axler-Blin, C., Vezinet-Brun, F., Rouzioux, C., Rozenbaum, W. and Montagnier, L. *Science*, **220**, 868-871 (1983).

Bernhard, W. *Cancer Res.*, **20**, 712-727, (1960).

Bolognesi, D.P. *Nature*, **340**, 431-432 (1989).

Bolognesi, D.P., Montelaro, R.C., Frank, H. and Schäfer, W. *Science*, **199**, 183-186 (1978).

Bosch, M.L., Earl, P.L., Fargnoli, K., Picciafuoco, S., Giombini, F., Wong Staal, F. and Franchini, G. *Science* **244**, 694-697 (1989).

Bouillant, A.M.P., Becker, S.A.W.E. *J. Natl. Cancer Inst.* **72**, 1075-1084 (1984).

Chrystie, I.L. and Almeida, J.D. *J. Med. Virol.* **27**, 188-195 (1989).

Clayton, L.K., Sieh, M., Pious, D.A. and Reinherz, E. L. *Nature* **339**, 548-551 (1989).

Cullen, B.R. and Greene, W.C. *Cell* **58**, 423-426 (1989).

Dahlberg, J.E. in 'Immunodeficiency Disorders and Retroviruses.', (Perk, K. ed.), Advances in Veterinary Science and Comparative Medicine, Vol. 32, pp. 1-35, Academic Press, London (1988).

Dalgleish, A.G., Beverly, P.C.L., Clapham, P.R., Crawford, D.H., Greaves, M.F. and Weiss, R.A. *Nature* **312**, 763-766 (1984).

Delchambre, M., Gheysen, D., Thines, D., Thiriart, C., Jacobs, E., Verdin, E., Horth, M., Burny, A. and Bex, F. *EMBO J.* **8**, 2653–2660 (1989).

Desrosiers, R.C., Daniel, M.D. and Li, Y. *AIDS Res. Hum. Retrovir.* **5**, 465–473 (1989).

Filice, G., Carnevale, G., Lanzarini, P., Orsolini, P., Soldini, L. and Cereda, P.M. *Microbiologica* **10**, 209–216 (1987).

Fine, D. and Schochetman, G. *Cancer Res.* **38**, 3123–3139 (1978).

Foley, J.D. and van Dam, A., 'Fundamentals of Interactive Computer Graphics', Massachusetts, Addison Wesley Publishing, Co. (1982).

Frank, H. 'Animal Virus Structure', (Nermut, M.V. and Steven, A.C. eds.), pp 295–304, Elsevier, Amsterdam. New York. Oxford (1987).

Frank, H., Schwarz, H., Graf, Th. and Schäfer, W. Z. *Naturforsch.* **33c**, 124–138 (1978).

Gallo, R., Wong Staal, F., Montagnier, L., Haseltine, W.A. and Yoshida, M. *Nature* **333**, 504 (1988).

Gallo, R.C., Salahuddin, S.Z., Popovic, M., Shearer, G.M., Kaplan, M., Haynes, B.F., Palker, T.J., Redfield, R., Oleske, J., Safai, B., White, G., Foster, P. and Markham, P. D. *Science* **224**, 500–503 (1984).

Gelderblom, H and Frank, H. 'Animal Virus Structure', (Nermut, M.V. and Steven, A.C. eds.), pp. 305–311 Elsevier, Amsterdam New York Oxford (1987).

Gelderblom, H., Özel, M. and Pauli, G. *Bundesgesundhbl.* **28**, 161–171 (1985a).

Gelderblom, H., Reupke, H. and Pauli, G. *Lancet* **11**, 1016–1017 (1985b).

Gelderblom, H., Reupke, H., Winkel, T., Kunze, R. and Pauli, G. Z. *Naturforsch.* **42c**, 1328–1334 (1987a).

Gelderblom, H.R., Hausmann, E., zel, M., Pauli, G. and Koch, M. *Virology* **156**, 171–176 (1987b).

Gelderblom, H.R., Özel, M. and Pauli G. *Arch. Virol.* **106**, 1–13 (1989).

Gelderblom, H.R., Özel, M., Hausmann, E.H.S., Winkel, T., Pauli, G., and Koch, M.A. *Micron Microscopica* **19**, 41–60 (1988).

Gendelman, H.E., Narayan, O., Kennedy-Stoskopf, S., Kennedy, P.G.E., Ghotbi, Z., Clements, J.E., Stanley, J. and Pezeshkpour, G. *J. Virol.* **58**, 67–74 (1986).

Gendelman, H.E., Orenstein, J.M., Baca, L.M., Weiser, B., Burger, H., Kalter, D.C. and Meltzer, M.S. *AIDS* **8**, 475–496 (1989).

Georgsson, G., Plsson, P.A. and P tursson, G. *Modern Pathol. AIDS* **7**, (in press) (1989).

Gheysen, D., Jacobs, E., de Foresta, F., Thiriart, C., Francotte, M., Thines, D. and De Wilde, M. *Cell* **59**, 103–112 (1989).

Göttlinger, H.G.,Sodroski, J.G. and Haseltine, W.A. *Proc. Natl. Acad. Sci. USA* **86**, 5781–5785 (1989).

Gonda, M.A. *J. Electron Microsc. Tech.* **8**, 17–40, (1988).

Gonda, M.A., Braun, M.J., Carter, S.G., Kost, T.A., Bess Jr, J.W., Arthur, L.O. and Van Der Maaten, M.J. *Nature* **330**, 388–391 (1987).

Gonda, M.A., Wong Staal, F., Gallo, R.C., *et al. Science* **227**, 173–177 (1985).

Goto, T., Harada, S., Yamamoto, N. and Nakai, M. *Arch. Virol.* **102**, 29–38 (1988).

Grewe C., Beck A. and Gelderblom H. R. *J. AIDS* in press (1990).

Grief, C., Hockley, D.J., Fromholc, C.E. and Kitchin, P.A. *J. Gen. Virol.* **70**, 2215–2219 (1989).

Haase, A.T. *Nature*, **322**, 130–136 (1989).

Haseltine, W. A. *J. AIDS* **1**, 217–240 (1988).

Haseltine, W.A. *J. AIDS* **2**, 311–334 (1989).

Haseltine, W.A., Terwilliger, E.F., Rosen, C.A., and Sodroski, J. G. in 'Retrovirus biology and human diseases.' (Gallo, R.C. and Wong Staal, F., eds),pp. 241-284, Marcel Dekker, Inc., New York and Basel (1989).

Henderson, L.E., Sowder, R., Copeland, T.D., Oroszlan, S., Arthur, L.O., Robey, W.G. and Fischinger, P.J. *J. Virol.* **61**, 629-632 (1987).

Henderson, L.E., Sowdner, R.C., Copeland, T.D., Benveniste, R.E. and Oroszlan, S. Science **241**, 199-201 (1988).

Hockley, D.J., Wood, R.D., Jacobs, J.P. and Garrett, A.J. J. Gen. Virol. **69**, 2455-2469 (1988).

Höglund, S., Öfverstedt, L. -G., Nilsson, Å., Özel, M., Winkel, T., Skoglund, U. and Gelderblom, H. in 'Retroviral Proteases: Control of Maturation and Morphogenesis', (Pearl,L.H. ed.), Macmillans, London, (1990).

Hoxie, J.A., Fitzharris, T.P., Youngbar, P.R., Matthews, D.M., Rackowski, J.L. and Radka, S.F. Hum. Immunol. **18**, 39-52 (1987).

Jacobs, E., Gheysen, D., Thines, D., Francotte, M. and de Wilde, M. Gene **79**, 71-81 (1989).

Kannagi, M., Kiyotaki, M., King, N.W., Lord, C.I. and Letvin, N.L. J. Virol. **61**, 1421-1426 (1987).

Katsumoto, T., Hattori, N. and Kurimura, T. Intervirol. **27**, 148-153 (1987).

Klatzman, D., Champagne, E., Chamaret, S., Gruest, J., Guetard, D., Hercend, T., Gluckman, J. C. and Montagnier, L. Nature **312**, 767-768 (1984).

Ladhoff, A M., Scholz, D., Rosenthal, S. and Rosenthal, H.A. Z. Klin. Med. **41**, 2209-2214(1986).

Lanzavecchia, A., Roosnek, E., Gregory, T., Berman, P. and Abrignani, S. *Nature* **334**, 530-532 (1988).

Lasky, L.A., Nakamura, G., Smith, D.H., Fennie, C., Shimasaki, C., Patzer, E., Berman, P., Gregory, T. and Capon, D.J. Cell 50, 975-985 (1987).

Lecatsas, G., Gravell, M. and Sever, J.L. *Proc. Soc. Exp. Biol. Med.* **177**, 495-498 (1984).

Leis, J., Baltimore, D., Bishop, J.M., Coffin, J., Fleissner, E., Goff, S.P., Oroszlan, S., Robinson, H., Skalka, A.M., Temin, H.M. and Vogt, V. *J. Virol.* **62**, 1808-1809 (1988).

Lever, A., Göttlinger, H., Haseltine, W. and Sodroski, J. *J. Virol.* **63**, 4085-4087 (1989).

Levy, J.A., Hoffmann, A.D., Kramer, S.M., Landis, J.A., Shimabukuro, J.M. and Oshiro, L.S. *Science*, **225** 840-842 (1984).

Levy, J.A., *JAMA*, **261**, 2997-3006 (1989).

Levy, J.A. *Nature*, **333**, 519-522 (1988).

Markham, R., Frey, S. and Hills, G.J. *Virology* **20**, 88-102 (1963).

Marsh M. and Dalgleish A. *Immunology Today* 8-12, 369-371 (1987).

Marsh, M. and Helenius, A. in 'Advances in Virus Research' **36**, (Maramorosch, K., Murphey, F.A. and Shatkin, A.J. ed.) pp. 107-158, Academic Press, San Diego, (1989).

Marx, P.A., Munn, R.J. and Joy, K.I. *Lab. Invest.* **58**, 112-118 (1988).

Matthews, R.E.F. 'Classification and nomenclature of viruses. Fourth report of the International Committee on Taxonomy of Viruses.', Karger, Basel (1982).

McClure, M., Marsh, M. and Weiss, R. *EMBO J.* **7**, 513-518 (1988).

McCune, J.M., Rabin, L.B., Feinberg, M.B., Liebermann, M., Kosek, J.C., Reyes, G.R. and Weissmann, I.L. *Cell* **53**, 55-67 (1988).

Modrow, S., Hahn, B.H., Shaw, G.M., Gallo, R.C., Wong Staal, F. and Wolf, H. *J. Virol.* **61**, 570-578 (1987).

Montagnier, L., Chermann, J.C., Barré-Sinoussi, F., Chamaret, S., Gruest, J., Nugeyre, M.T., Rey, F., Dauguet, C., Axler-Blin, C., Vézinet-Brun, F., Rouzioux, C., Saimot, G. A., Rozenbaum, W., Gluckmann, J.C., Klatzmann, D., Vilmer, E., Griscelli, G., Foyer-Gazengel. C. and Brunet, J.B. in 'Human T cell Leukemia/Lymphoma Virus', (R.C. Gallo, M.E. and L. Gross ed.), pp. 363-379, Cold Spring Harbor Laboratory, (1984).

Montelaro, R.C., Robey, W.G., West, M.D., Issel, C.J. and Fischinger, P.J. *J. Gen. Virol.* **69**, 1711-1717 (1988).

Munn, R., Marx, P.A., Yamamoto, J.K. and Gardner, M.B. Lab. Invest. 53, 194-199 (1985).

Nakai, M., Goto, T. and Imura, S. J. Electron Microsc. Tech. 12, 95-100 (1989).

Narayan, O. and Clements J.E. J. Gen. Virol. 70, 1617-1639 (1989).

Oroszlan, S. and Luftig, R.B. in 'Current Topics in Microbiology and Immunology' **157**, (1990).

Overton, H.A., Fujii, Y., Price, I.R. and Jones, I.M. *Virology* **170**, 107-116 (1989).

Özel, M., Pauli, G., and Gelderblom, H.R. *Arch. Virol.* **100**, 255-266 (1988).

Palmer, E. and Goldsmith, C.S. *J. Electron Microsc. Tech.* **8**, 3-15 (1988).

Palmer, E., Martin, M.L., Goldsmith, C. and Switzer, W. *J. gen. Virol.* **69**, 1425-1429 (1988).

Parekh, B., Issel, C.J. and Montelaro, R.C. *Virology.* **107**, 520-525 (1980).

Pauli, G., Rohde, G. and Harms, E. *Arch. Virol.* **58**, 61-64 (1978).

Pauza, C.D. and Price, T.M. *J. Cell Biol.* **107**, 959-968 (1988).

Pinter, A. and Fleissner, E. *Virology* **83**, 417-422 (1977).

Pinter, A., Honnen, W.J., Tilley, S.A., Bona, C., Zaghouani, H., Gorny, M.K. and Zolla-Pazner, S. *J. Virol.* **63**, 2674-2679 (1989).

Putney, S.D. and Montelaro, R.C. in 'Immunochemistry of Viruses II' (Neurath and van Regenmortel eds.), Elsevier Biochemical Press, in press (1989).

Rein, A., McClure, M.R., Rice, N.R., Luftig, R.B. and Schultz, A.M. *Proc. Natl. Acad. Sci. USA* **83**, 7246-7250 (1986).

Ribas, T., Hase, E., Hunter, E., Fritz, D., Khan, N. and Burke, D. IV International Conference on AIDS, Stockholm, Abstract No 1025 (1988).

Roberts M.M. and Oroszlan S. *Biochem. Biophys. Res. Comm.* **160**, 486-494 (1989).

Rosenberg, Z.F. and Fauci, A.S. *Clin. Immunol. Immunopathol.* **50**, 149-156 (1989). Sarkar, N.H. in 'Animal Virus Structure' (M.V. Nermut and A.C. Steven eds.), pp. 257-272, Elsevier, Amsterdam. New York. Oxford, (1987).

Schawaller, M., Smith, G.E., Skehel, J.J. and Wiley, D.C. *Virology* **172**, 367-369 (1989).

Schidlovsky, G., Shibley, G.P., Benton, C.V. and Elser, J.E. *J. Natl. Cancer Inst.* **61**, 91-95 (1978).

Schneider, J. and Hunsmann, G. *Int. J. Cancer* **22**, 204-213 (1978).

Schneider, J., Jurkiewicz, E., Wendler, I., Jentsch, K.D., Bayer, H., Desrosiers, R.C., Gelderblom, H. and Hunsmann, G. in 'Viruses and Human Cancer' (R.C. Gallo, W. Haseltine, G. Klein and H. zur Hausen eds.) pp 319-332, Alan R. Liss, Inc., New York, (1987).

Schneider, J., Kaaden, O., Copeland, T.D., Oroszlan, S. and Hunsmann, G. *J. Gen. Virol.* **37**, 2533-2538 (1986).

Siliciano, R.F., Lawton, T., Knall, C., Karr, R.W., Berman, P., Gregory, T. and Reinherz, E.L. *Cell* **54**, 561-575 (1988).

Sölder, B.M., Schulz, T.F., Hengster, P., Löwer, J., Larcher, C., Bitterlich, G., Kurth, R., Wachter, H. and Dierich, M.P. *Immunol. Lett.* **22**, 135–146 (1989).

Stein, B.S., Gowda, S.D., Lifson, J.D., Penhallow, R.C., Bensch, K.G. and Engleman, E.G. *Cell* **49**, 659–668 (1987).

Takahashi, I., Takama, M., Ladhoff, A.M. and Scholz, D. *J. AIDS* **2**, 136–140 (1989).

Veronese, F.M., Copeland, T.D., Oroszlan, S., Gallo, R.C. and Sarngadharan, M.G. *J. Virol.* **62**, 795–801 (1988).

Weiland, F. and Bruns, M. *Arch. Virol.* **64**, 277–285 (1980).

Weiss R., Teich N., Varmus H. and Coffin J. 'RNA tumor viruses: molecular biology of tumor viruses.', 2nd edn. Cold Spring Harbour Laboratory, New York (1984).

Weiss, R.A. *Nature* **333**, 497–498 (1988).

Index